COLORADO

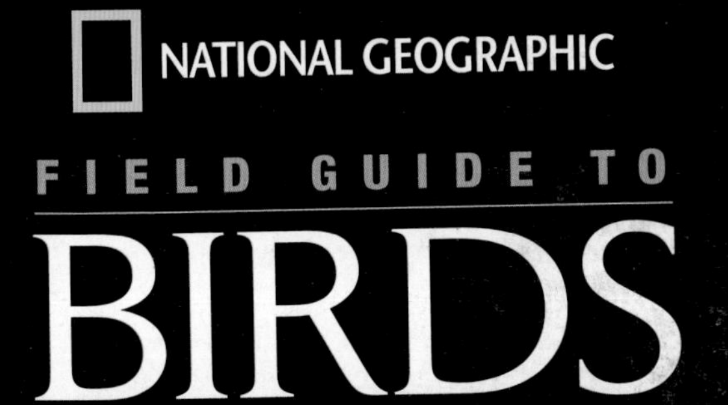

COLORADO

Edited by JONATHAN ALDERFER

National Geographic
Washington, D.C.

Introduction

Colorado has some of North America's most enjoyable birdwatching. The diversity of habitats—from alpine tundra, to shortgrass prairie, to coniferous forest—places Colorado at the crossroads of the bird world. Roughly 480 species have been recorded, of which about 350 can be expected every year, and most can be visited within a couple hours' drive from Denver and the rest of Colorado's Front Range.

The Black-billed Magpie and Red-tailed Hawk can be found in a variety of habitats, but to see White-tailed Ptarmigan and Brown-capped Rosy-Finch, you'll need to visit the alpine tundra—accessible from Denver at Rocky Mountain National Park, Mount Evans, and Guanella Pass. What about Mountain Plover, Chestnut-collared and McCown's Longspur? Try Pawnee National Grasslands, an hour north of Denver. But also be prepared to travel. Southeastern Colorado has specialties like Lesser Prairie-Chicken and Greater Roadrunner. The West Slope harbors the Gunnison Sage-Grouse and Gray Vireo, and its coniferous forests offer the Pine Grosbeak and Three-toed Woodpecker. The Greater Prairie-Chicken and Red-headed Woodpecker are confined to the grasslands or riparian corridors in Colorado's northeastern corner.

During spring in Colorado, lekking grouse bring hundreds of birders. In summer, resident species are augmented by breeders, most of which are easy to see. During fall migration, anything can show up. And winter offers your best chance to see all the Rosy-Finches and an all-white White-tailed Ptarmigan. Whether you live in the state or are visiting, you'll see why Colorado is one of North America's premier birding destinations.

Chris Wood
Cornell Laboratory of Ornithology

Frontispiece: White-tailed Ptarmigan in summer
Photo by Chris Wood

CONTENTS

SELECTED BIRDING SITES OF COLORADO

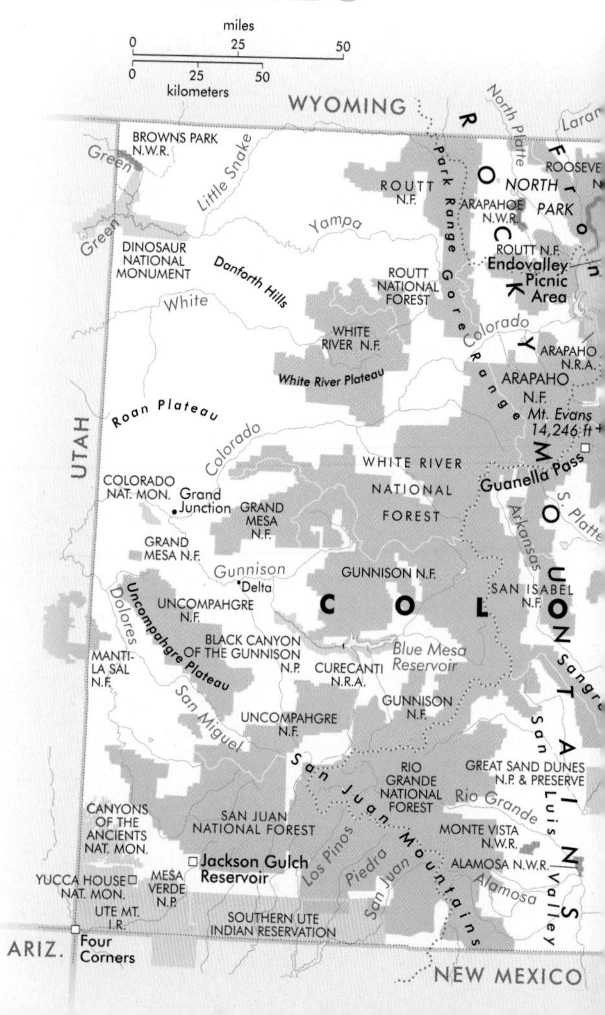

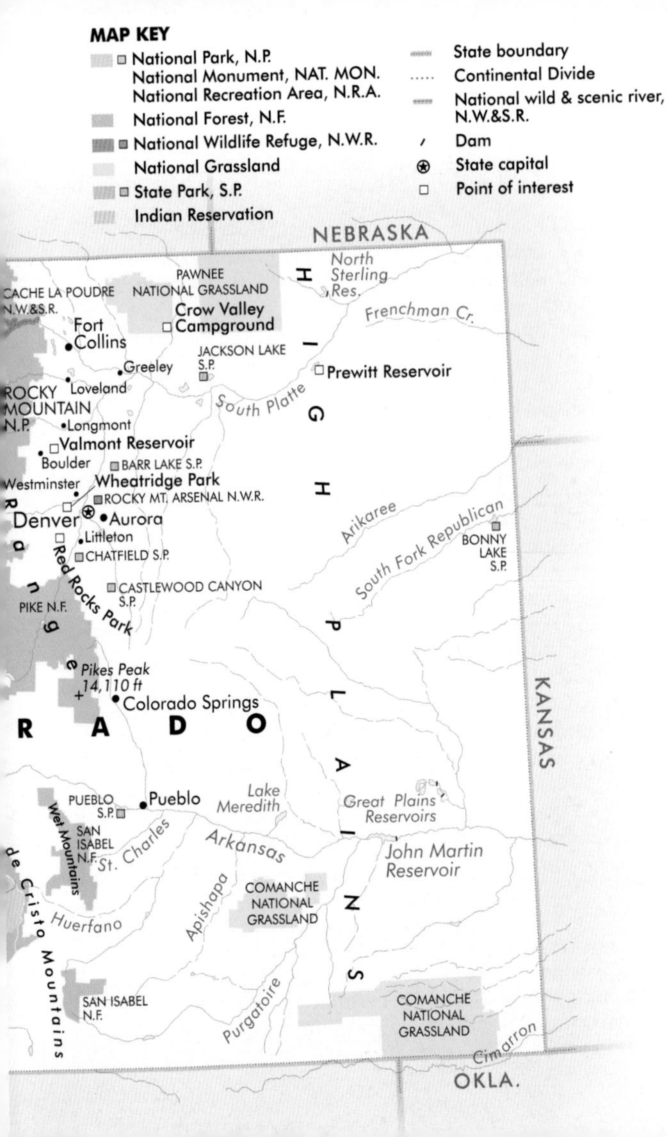
MAP KEY
National Park, N.P.
National Monument, NAT. MON.
National Recreation Area, N.R.A.
National Forest, N.F.
National Wildlife Refuge, N.W.R.
National Grassland
State Park, S.P.
Indian Reservation
State boundary
Continental Divide
National wild & scenic river, N.W.&S.R.
Dam
State capital
Point of interest
NEBRASKA
KANSAS
OKLA.
HIGH PLAINS
RADO
North Sterling Res.
Frenchman Cr.
PAWNEE NATIONAL GRASSLAND
CACHE LA POUDRE N.W.&S.R.
Crow Valley Campground
Fort Collins
Greeley
JACKSON LAKE S.P.
Prewitt Reservoir
Loveland
ROCKY MOUNTAIN N.P.
South Platte
Longmont
Valmont Reservoir
Boulder
BARR LAKE S.P.
Westminster
Wheatridge Park
ROCKY MT. ARSENAL N.W.R.
Denver
Aurora
Littleton
CHATFIELD S.P.
Arikaree
South Fork Republican
BONNY LAKE S.P.
Range
Red Rocks Park
CASTLEWOOD CANYON S.P.
PIKE N.F.
Pikes Peak 14,110 ft
Colorado Springs
PUEBLO S.P.
Pueblo
Lake Meredith
Great Plains Reservoirs
Wet Mountains
SAN ISABEL N.F.
St. Charles
Arkansas
John Martin Reservoir
Apishapa
COMANCHE NATIONAL GRASSLAND
Sangre de Cristo Mountains
Huerfano
SAN ISABEL N.F.
Purgatoire
COMANCHE NATIONAL GRASSLAND
Cimarron

Looking at Birds

What do the artist and the nature lover share? A passion for being attuned to the world and all of its complexity, via the senses. Every time you go out into the natural world, or even view it through a window, you have another opportunity to see what is there. And the more you know what you are looking at, the more you see.

Even if you are not yet a committed birder, it makes sense to take a field guide with you when you go out for a walk or a hike. Looking for and identifying birds will sharpen and heighten your perceptions, and intensify your experience. And you'll find that you notice everything else more acutely—the terrain, the season, the weather, the plant life, other animal life.

Birds are beautiful, complex animals that live everywhere around us in our towns and cities and in distant places we dream of visiting. Here in North America more than 900 species have been recorded—from abundant commoners to the rare and exotic. A comprehensive field reference like the *National Geographic Field Guide to the Birds of North America* is essential for understanding that big picture. If you are taking a summer hike in the Colorado Rockies, however, you may want something simpler: a guide to the birds you are most likely to see, which slips easily into a pocket.

This photographic guide is designed to provide an introduction to the common birds—and some of the specialty birds—you might encounter in Colorado: how to identify them, how they behave, and where to find them, with specific locations.

Discovery, observation, and identification of birds is exciting, whether you are novice or expert. I know that every time I go out to look at birds, I see more clearly and have a greater appreciation for the natural world and my own place in it.

Jonathan Alderfer
Editor

How to Use this Book

National Geographic Field Guide to Birds: Colorado is designed to help beginning and practiced birders alike identify birds in the field and introduce them to the region's varied birdlife. The book is organized by bird families, following the order in the *Check-list of North American Birds*, by the American Ornithologists' Union. Families share structural characteristics, and by learning these shared characteristics early, birders can establish a basis for a lifetime of identifying birds and related family members with great accuracy—sometimes merely at a glance. (For quick reference in the field, use the color and alphabetical indexes at the back of this book.)

A family may have one member or dozens of members, or species. In this book each family is identified by its common name in English along the right-hand border of each spread. Each species is also identified in English, with its Latin genus and species—its scientific name—found directly underneath. One species is featured in each entry.

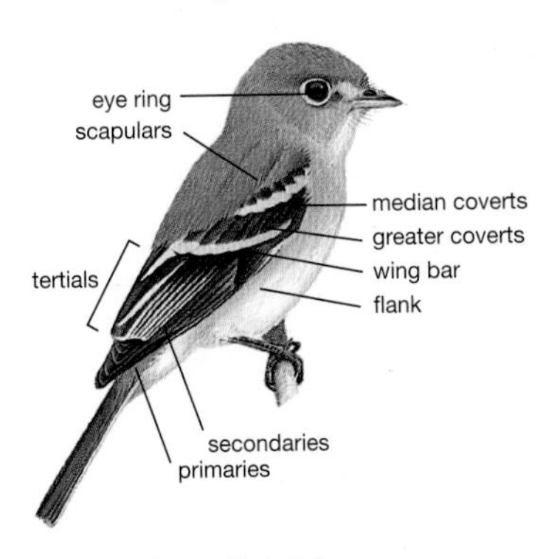

Least Flycatcher

Lark Sparrow

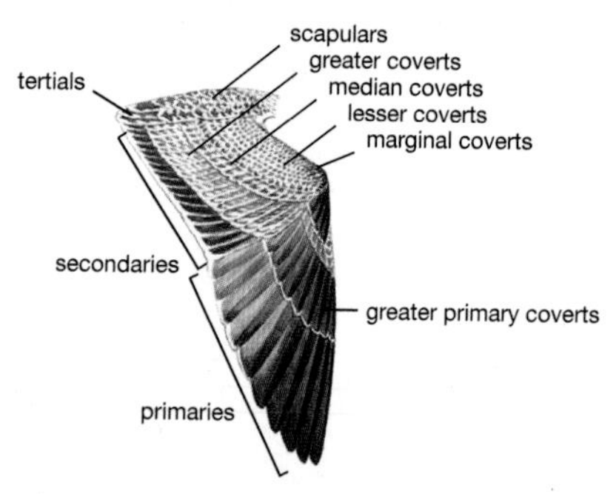

Great Black-backed Gull, upper wing

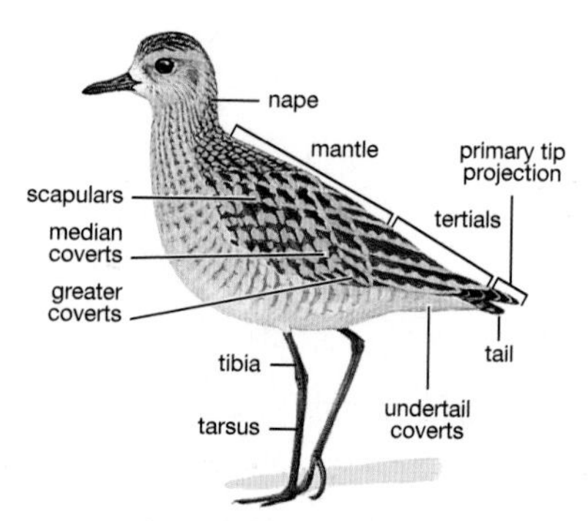

Pacific Gold en-Plover

An entry begins with **Field Marks**, the physical clues used to quickly identify a bird, such as body shape and size, bill length, and plumage color or pattern. A bird's body parts yield vital clues to identification, so a birder needs to become familiar with them early on. After the first glance at body type, take note of the head shape and markings, such as stripes, eye rings, and crown markings. Bill shape and color are important as well. Note body and wing details: wing bars, color of and pattern of wing feathers at rest, and shape and markings when extended in flight. Tail shape, length, color, and banding may play a big part, too. At left are diagrams detailing the various parts of a bird—its topography—labeled with the term likely to be found in the text of this book.

The main body of each entry is divided into three categories: Behavior, Habitat, and Local Sites. The **Behavior** section details certain characteristics to look or listen for in the field. Often a bird's behavioral characteristics are very closely related to its body type and field marks, such as in the case of woodpeckers, whose stiff tails, strong legs, and sharp claws enable them to spend most of their lives in an upright position, braced against a tree trunk. The **Habitat** section describes areas that are most likely to support the featured species. Preferred nesting locations of breeding birds are also included in many cases. The **Local Sites** section recommends specific refuges or parks where the featured bird is likely to be found. A section called **Field Notes** finishes each entry, and includes information such as plumage variations within a species; or it may introduce another species with which the featured bird is frequently confused. In either case, an additional drawing may be included to aid in identification.

Finally, a caption under each of the photographs gives the season of the plumage pictured, as well as the age and sex of the bird, if discernable. A key to using this informative guide and its range maps follows on the next two pages.

Reading the Spread

❶ **Photograph:** Shows bird in habitat. May be female or male, adult or juvenile. Plumage may be breeding, molting, nonbreeding, or year-round.

❷ **Caption:** Defines the featured bird's plumage, age, and sometimes sex, as seen in the picture.

❸ **Heading:** Beneath the common name find the Latin, or scientific, name. Beside it is the bird's length (L), and sometimes its wingspan (WS). Wingspan is given with birds often seen in flight. Female measurements are given if notably different from male.

❹ **Field Marks:** Gives basic facts for field identification: markings, head and bill shape, and body size.

❺ **Band:** Gives the common name of the bird's family.

❻ **Range Map:** Shows year-round range in purple, breeding range in red, winter range in blue. Areas through which species are likely to migrate are shown in green.

❼ **Behavior:** A step beyond **Field Marks,** gives clues to identifying a bird by its habits, such as feeding, flight pattern, courtship, nest-building, or songs and calls.

❽ **Habitat:** Reveals the area a species is most likely to inhabit, such as forests, marshes, grasslands, or urban areas. May include preferred nesting sites.

❾ **Local Sites:** Details local spots to look for the given species.

❿ **Field Notes:** A special entry that may give a unique point of identification, compare two species of the same family, compare two species from different families that are easily confused, or focus on a historical or conservation fact.

Reading the Maps

On each map of Colorado, range boundaries are drawn beyond which the species is not regularly seen. Nearly every species will be rare at the edges of its range. The sample map shown below explains the colors and symbols used on each map. Ranges continually expand and contract, so the map is a tool, not a rule. Range information is based on actual sightings and therefore depends upon the number of knowledgeable and active birders in each area.

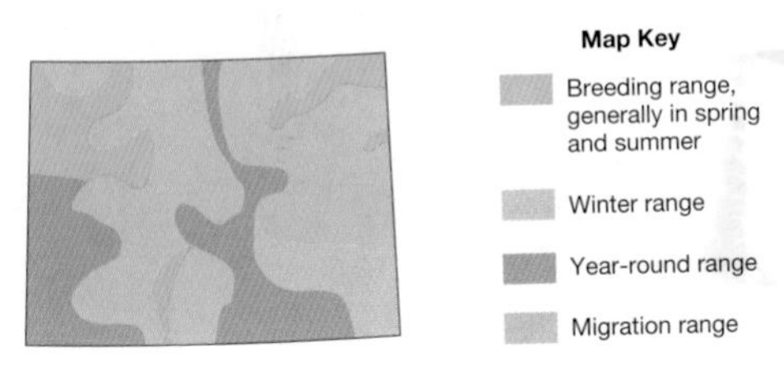

Sample Map: Spotted Towhee

Reading the Indexes

There are two indexes at the back of this book. The first is a **Color Index** (p. 262), created to help birders quickly find an entry by noting its color in the field. In this index, birds are labeled by their predominant color: Mostly White, Mostly Black, etc. Note that a bird may have a head of a different color than its label states. That's because its body—the part most noticeable in the field—is the color labeled.

The **Alphabetical Index** (p. 266) is organized by the bird's common name. Next to each entry is a check-off box. Most birders make lists of the birds they see. Some keep several lists, perhaps one of birds in a certain area and another of all the birds they've ever seen—a life list. Such lists enable birders to look back and remember their first sighting of a Lazuli Bunting or an American Kestrel.

Year-round | Adult

CANADA GOOSE

Branta canadensis L 45" (114 cm)

FIELD MARKS

Black head and neck marked with distinctive white chin strap

In flight shows large, dark wings; white undertail coverts and a long protruding neck

Variable pale gray-brown breast

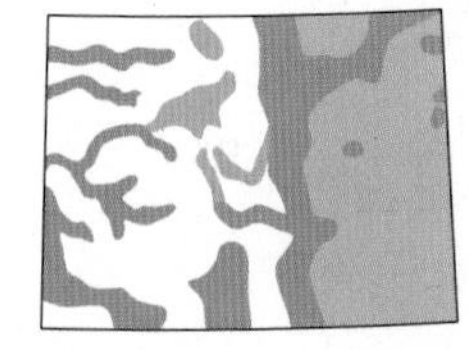

Behavior

A common, familiar goose in winter; best known for migrating in large V-formation. Its distinctive musical call of *honk-a-lonk* makes it easy to identify, even without seeing it. Like some other members of its family, the Canada Goose finds a mate and remains monogamous for life. Family groups tend to stay together through the winter.

Habitat

Prefers ponds and lakes, also in cultivated fields. It has adapted successfully to habitats such as golf courses and farms, sometimes chasing off other nesting waterbirds.

Local Sites

Abundant migrant throughout Colorado, and common breeder on wetlands and golf courses. Largest concentrations winter near metropolitan areas along the Front Range.

FIELD NOTES In 2004, the American Ornithologists' Union split the Canada Goose into two species, making the Cackling Goose a separate species, *Branta hutchinsii* (inset). The Cackling Goose, a fairly common migrant and uncommon winter visitor in Colorado, is smaller (L25-30") with shorter neck and stubby bill.

Breeding | Adult male

GADWALL

Anas strepera L 20" (51 cm)

FIELD MARKS

Male mostly gray; brownish head and back; black tail coverts; chestnut wing patches

Female mottled brown overall; dark upper mandible has distinctive orange sides

White speculum shows in flight

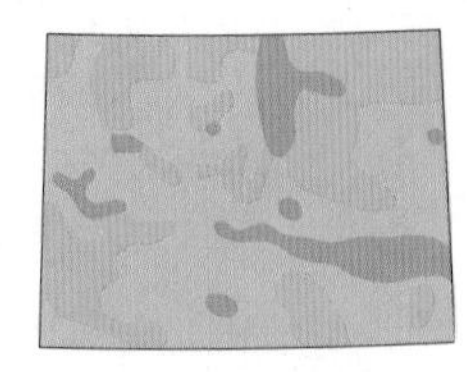

Behavior

Feeds primarily on aquatic vegetation, insects, and other invertebrates, in shallow water. Found in pairs or small groups, foraging with its head submerged, but without tipping up like many other dabbling ducks. Also known to dive for its food in deeper waters. Walks well on land, and may be seen foraging in fields for grains and seeds. Female builds nest of weeds and grasses on dry land near water. Female's call is a descending series of loud quacks; male sometimes emits a shrill, whistled note.

Habitat

Resides primarily in freshwater habitats, especially those with dense vegetation.

Local Sites

The Gadwall is widespread year-round in most locations with open water. Check any pond or lake from late fall to early spring, when it is most easily seen.

FIELD NOTES Another common duck found in Colorado, the American Wigeon, *Anas americana* (inset: adult male), also has white on its wing. Note that the white on a wigeon is on the upper wing, not on the secondaries.

Breeding | Adult male

MALLARD

Anas platyrhynchos L 23" (58 cm)

FIELD MARKS

Male has metallic green head and neck; white collar; chestnut breast

Female mottled brown overall; orange bill marked with black

Both sexes have bright blue speculum bordered in white; white tail and underwings

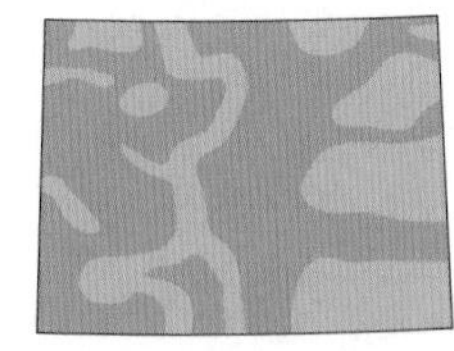

Behavior

A dabbler, the Mallard feeds by tipping up in shallow water and plucking seeds, grasses, or invertebrates from the bottom. Also picks insects from the water's surface. The courtship ritual of the Mallard consists of the male pumping his head, dipping his bill, and rearing up in the water to exaggerate his size. A female signals consent by duplicating the male's head-pumping. Nests on the ground in concealing vegetation. Listen for the female Mallard's loud, rasping quack.

Habitat

This widespread species occurs wherever shallow fresh water is to be found, from pristine lakes to urban ponds.

Local Sites

Mallards cope well with man-made habitats and reside all year in ponds, streams, and city fountains throughout Colorado.

FIELD NOTES The female Mallard (inset) is mottled brown overall. Its head is distinctly grayer than its body, and it displays a dark eye stripe. Between June and September, the male resembles the female, but retains his yellowish bill, while the female's bill is orange with a dark center.

Breeding | Adult male

CINNAMON TEAL

Anas cyanoptera L 16" (41 cm)

FIELD MARKS

Male has cinnamon head, neck, and underparts

Female is mottled brown overall

Red-orange eye; long, spatulate, blackish bill

Bright blue upperwing coverts

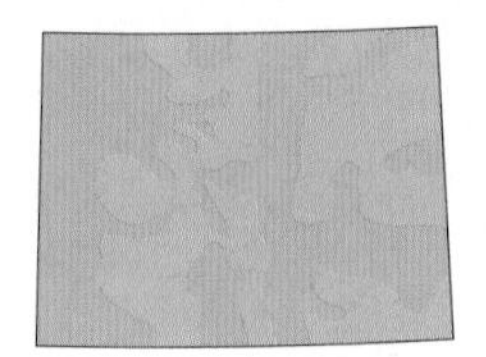

Behavior

Small but powerful, like other dabblers, the Cinnamon Teal takes flight by leaping directly into the air. Its omnivorous diet changes according to seasonal availability and particular needs during breeding, molting, and migrating. Teals typically pick insects from the water's surface or pluck grasses and invertebrates from the bottom. Builds nest on ground near edges of wetlands, often below concealing dead and matted vegetation. The female gets into the nest through tunnels in the vegetation.

Habitat

Prefers freshwater marshes, ponds, and lakes.

Local Sites

Fairly common to uncommon breeder on small lakes and ponds throughout Colorado. More commonly seen as a migrant.

FIELD NOTES The male Blue-winged Teal, *Anas discors* (inset), has a distinctive white crescent on its face, but the female is very similar to the female Cinnamon Teal. In flight, the Cinnamon Teal's wing pattern matches that of the Blue-winged Teal, which is a common migrant and breeder in Colorado.

Breeding | Adult male

NORTHERN SHOVELER

Anas clypeata L 19" (48 cm)

FIELD MARKS

Large spatulate bill

Male has green head, white breast, chestnut belly; white facial crescent in fall

Female mottled brown overall, grayish orange bill

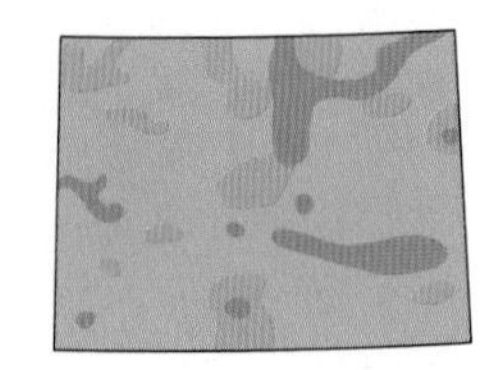

Behavior

This dabbling duck is equipped with a unique bill that has comblike bristles to strain plankton and insects from water. Forages while swimming, its bill submerged or skimming the surface for aquatic seeds and plants. In shallow water, the Northern Shoveler sieves through muddy bottoms for small crustaceans and mollusks. Nests in short, dense grasses close to a body of water. Listen for the female's deep, descending *whack-whack-whack-wak-waa,* and during breeding season for the male's hoarse, unmusical *tuk-tuk-tuk.*

Habitat

Favors small, shallow lakes and ponds, freshwater and saline marshes, and other smaller bodies of water densely bordered by emergent vegetation.

Local Sites

Shovelers are fairly widespread throughout Colorado; they are most widely seen in migration and during summer months. In winter, they keep more to the Front Range.

FIELD NOTES Another dabbler found throughout Colorado, the aptly named Northern Pintail, *Anas acuta* (inset, male), has a thin grayish bill, long pointed tail, long neck, and long slender wings.

Breeding | Adult male

GREEN-WINGED TEAL

Anas crecca L 14.5" (37 cm)

FIELD MARKS

Male's chestnut head has green ear patch faintly outlined in white

Female has mottled, dusky brown upperparts; white belly and undertail coverts

In flight, shows green speculum bordered above in buff

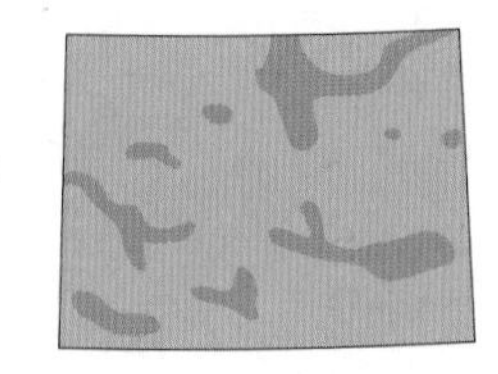

Behavior

An agile and fast-moving flier, this is the smallest species of ducks known as dabblers. A dabbler either feeds at the water's surface or upended, tail in the air and head submerged, to reach aquatic plants, seeds, and snails. The Green-winged has a specialized bill for filtering food from the mud. Travels in small flocks that synchronize their twists and turns in midair. The Green-winged hen emits a high, shrill *skee.*

Habitat

Found on shallow lakes and ponds, especially those with standing or floating vegetation. Also known to feed in inland agricultural and wooded areas.

Local Sites

Like many of Colorado's wintering waterfowl, flocks of Green-winged Teals converge along the Front Range, over rivers and lakes. They are also frequently seen in summer.

FIELD NOTES The female Green-winged Teal (inset) is mottled brown overall with a small, dark bill. She can be told from other female ducks by her largely white undertail coverts and green speculum, bordered above in buff.

Breeding | Adult male

REDHEAD

Aythya americana L 19" (48 cm)

FIELD MARKS

Male has light back and sides, black breast and tail, rufous head

Female is brown overall with a patch of white below the bill

Bill is blue-gray with white ring and black tip

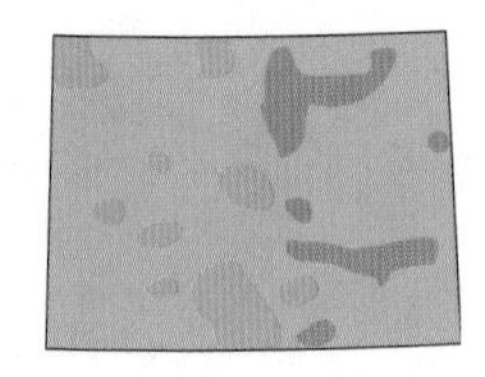

Behavior

Flocks congregate into "floating rafts" in winter, feeding mostly at dawn and dusk by diving for aquatic vegetation. Flies with strong wing beats in a V-formation. On breeding grounds, females deposit their eggs in the nests of other waterfowl species, sometimes even in American Bittern *(Botarus lentiginosus)* nests. Female's call is a rough, grating *squak.*

Habitat

Winters on large lakes and reservoirs, often mixed with other species.

Local Sites

Found in winter and migration on large bodies of water throughout Colorado. In summer, the Redhead is most easily found in mountain parks.

FIELD NOTES The closely related Canvasback, *Aythya valisineria* (inset: male, left; female, right), is often a victim of the Redhead's brood parasitism on their shared breeding grounds from the northern Great Plains to Alaska. The Canvasback male's head is a deeper maroon, his back and sides are white. The female has a dusky brownish gray body. Both have solid black bills, strongly sloping foreheads.

Breeding | Adult male

RING-NECKED DUCK

Aythya collaris L 17" (43 cm)

FIELD MARK

Male has black head, breast, back, and tail; pale gray sides

Female is brown with pale face patch, eye ring, and eye stripe

Peaked crown; blue-gray bill with white ring and black tip

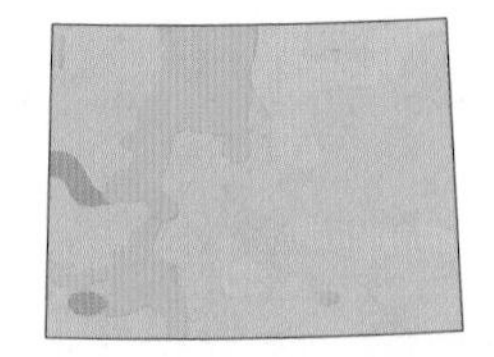

Behavior

An expert diver, the Ring-necked can feed as deep as 40 feet below the surface, but tends to remain in shallower waters. Small flocks can often be seen diving for plants, roots, and seeds. Unlike most other diving ducks, the Ring-necked springs into flight directly from water, and flies in loose flocks with rapid wing beats. Though often silent, the Ring-necked female sometimes gives a harsh, grating *deeer.*

Habitat

Inhabits freshwater marshes, woodland ponds, and small lakes. Often nests atop a floating raft of aquatic vegetation.

Local Sites

Ring-neckeds move to lower elevations in the mountains in winter. Observations suggest their numbers as breeders in the state have increased since the 1930s.

FIELD NOTES The Lesser Scaup, *Aythya affinis* (inset: breeding male, left; female, right) can be distinguished from the Ring-necked by its bluish gray bill. It is found in wetlands across Colorado, primarily in winter. The Lesser's larger cousin, the Greater Scaup, *Aythya marila,* is much less common; it has a more rounded crown and a larger bill.

Breeding | Adult male

COMMON GOLDENEYE

Bucephala clangula L 18.5" (47 cm)

FIELD MARKS

Triangular head brown in female; black with greenish tinge in male

Male has white patch between eye and bill; female has brown head

Male has black upperparts; female grayish

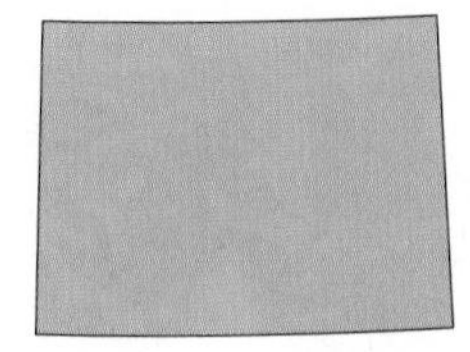

Behavior

A diving duck; often seen foraging in flocks. In flight, look for a distinctive white band on the secondaries of both the male and female, and listen for a whistling sound as they pass overhead.

Habitat

Prefers open lakes near woodlands where nest holes are available in large tree cavities. In winter, retires to lakes, rivers, and reservoirs. May sometimes use nest boxes or abandoned buildings for its nest, lining the depression with wood chips and down.

Local Sites

The Common Goldeneye is commonly seen in winter throughout Colorado where there is open water. They are easily seen along most of the state's major rivers, in and near the foothills.

FIELD NOTES The striking male Bufflehead, *Bucephala albeola* (inset), is commonly seen throughout Colorado, mostly in migration and in winter. The dark grayish females are also distinctive, with an oval-shaped white patch on the side of the head.

Nonbreeding | Adult male (left) and female

COMMON MERGANSER

Mergus merganser L 25" (64 cm)

FIELD MARKS

Breeding male has blackish green head and black scapulars

Female and nonbreeding male have chestnut head and gray back

Slim neck

Red bill is thin, hooked at tip

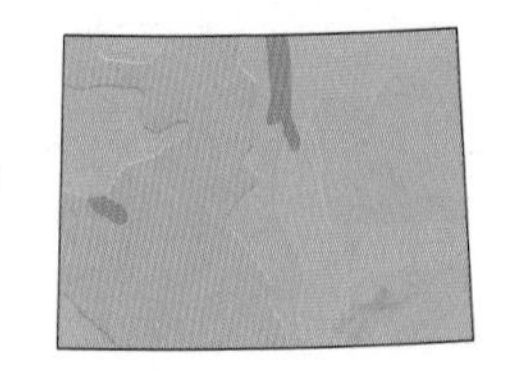

Behavior

Swiftly gives chase to small fish underwater. A long, thin, serrated bill helps it to catch fish, mollusks, crustaceans, and aquatic insects. Flies low with rapid wing beats, following the course of rivers and streams. Pairs form in late winter, before which this bird is most commonly found in single-gender flocks of 10 to 20 birds. Harsh croaks can be heard from the male; a loud, harsh *carr-carr* from the female.

Habitat

Prefers the still, open water of large lakes, but may also be found along rivers, particularly during summer. Nests in woodlands in tree cavities, and rock crevices near lakes and rivers.

Local Sites

Common Mergansers can be found on large bodies of water throughout most of Colorado, mostly in winter. Huge concentrations of several thousand may be seen at John Martin Reservoir in Bent County.

FIELD NOTES The less common female Red-breasted Merganser, *Mergus serrator,* is very similar to the female Common Merganser, but the Red-breasted's head and neck are paler, and lack the Common's distinct white chin.

Year-round | Adult male, displaying

GUNNISON SAGE-GROUSE

Centrocercus minimus Male L 22" (56 cm) Female L 18" (46 cm)

FIELD MARKS

Large size; mostly black belly; long tail

Male has featherless yellow air sacs on breast, inflated during displays

Very limited range

Behavior

Gunnison Sage-Grouse males gather each spring to perform an elaborate courtship display. Each male spreads his tail and rapidly inflates and deflates his air sacs, which emits a deep bubbling sound. At the same time, the long feathers on the back of his head are thrown forward, then flop back—behavior unique to the Gunnison, and one reason why it is now considered a separate species from the Greater Sage-Grouse.

Habitat

Prefers low vegetation dominated by sage, near rivers or other water sources. Courtship arenas, or leks, are in more open areas surrounded by sage.

Local Sites

An uncommon and declining species found in the intermountain valleys of southwestern Colorado. Look for it near dawn or dusk at the public Waunita Hot Springs lek east of Gunnison.

FIELD NOTES The Greater Sage-Grouse, *Centrocercus urophasianus* (inset: female, left; male, right), is found in northwestern Colorado. The two species do not overlap in distribution and are best identified by range. Also note Gunnison's smaller size, paler overall coloration, white-banded tail, and differences in breeding displays.

Winter | Adult

WHITE-TAILED PTARMIGAN

Lagopus leucura L 12.5" (32 cm)

FIELD MARKS

In winter, all white plumage

In summer, mostly mottled plumage

White tail and feet year-round

Only grouse likely to be observed on alpine tundra

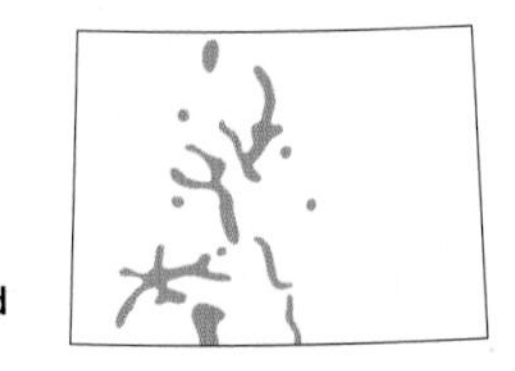

Behavior

The White-tailed Ptarmigan forages on a variety of buds, blossoms, flowers, and other herbaceous vegetation. Its camouflaged plumage allows it to forage leisurely, often allowing observers to approach closely. Though fairly quiet, emits soft clucking in the summer, as chicks and adults feed together. In late spring, the male White-tailed Ptarmigan gives loud cries, chatters and clucks as it proclaims its territory.

Habitat

Inhabits alpine tundra. In winter, birds move slightly south into areas with more willows. In summer they are often in very open areas with lush flower growth.

Local Sites

Found above timberline in most of Colorado's mountain ranges. Good summer locations for seeing ptarmigans include Rocky Mountain National Park and Mount Evans. In winter try Guanella Pass.

FIELD NOTES The mottled plumage of a White-tailed Ptarmigan in summer (inset) can make it especially hard to find. More than one observer has remarked that the best way to find one is to head up to the alpine tundra, sit down, and patiently eat lunch. When you see a rock move, you've found your ptarmigan!

Year-round | Adult male, displaying

BLUE GROUSE

Dendragapus obscurus L 16" (43 cm)

FIELD MARKS

Black tail with broad pale terminal band

Male largely dark dusky gray; female mottled gray and brown

Displaying males reveal inflatable neck sacs

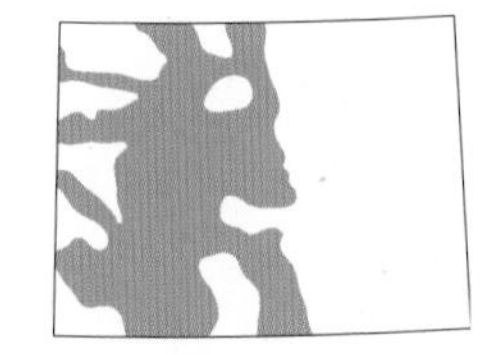

Behavior

Courting season in Colorado finds the Blue Grouse male inflating reddish neck sacs in morning and early evening displays. Unlike many of Colorado's grouse, the Blue Grouse does not form large communal leks, but perches on the ground and gives a low hoot. Such differences are leading to the imminent split of the Blue Grouse into two separate species: the interior "Dusky Grouse" and coastal "Sooty Grouse."

Habitat

The most likely grouse to be encountered in forests and woodlands, from Gambel's oak to mixed conifer-aspen. Some birds move from relatively open areas during breeding season to dense conifer forests in autumn.

Local Sites

Most easily seen in the western and southern mountains. Try the Black Canyon of the Gunnison National Park either early or late in the day.

FIELD NOTES Once more widespread, the Sharp-tailed Grouse, *Tympanuches phasianellus* (inset: displaying male), is now a year-round resident in northwestern Colorado and the extreme northeastern corner of the state. In spring, the male's courtship display includes foot stomping, tail shaking, and wing spreading.

Year-round | Adult male, displaying

LESSER PRAIRIE-CHICKEN

Tympanuchus pallidicinctus L 16" (41 cm)

FIELD MARKS

Heavily barred with brown, buff, and white

Male has yellow-orange eye combs and inflatable neck sacks

Sexes similar in plumage: difficult to tell apart away from leks

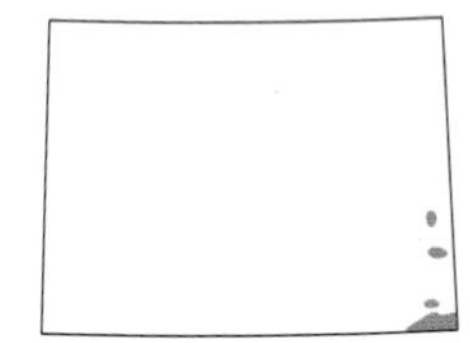

Behavior

In spring, males gather at communal display grounds, called leks, where they attempt to attract and mate with females by engaging in complex displays accompanied by shrill squawks and eerie cackles. Leks in Colorado rarely exceed 40 birds. Lesser Prairie-Chickens are declining throughout their range, including Colorado.

Habitat

Found in short-grass prairie in southeastern Colorado, favoring locations with healthy and diverse grasslands often interspersed with sand sage. Throughout most of its range prefers habitat with shinnery oak.

Local Sites

Most observations of Lesser Prairie-Chickens occur southeast of Springfield on the Comanche National Grasslands. April offers the best chance of seeing the unforgettable display of this declining species.

FIELD NOTES Replacing the Lesser Prairie-Chicken in northeastern Colorado, the Greater Prairie-Chicken, *Tympanuchus cupido* (inset: displaying male), is very similar in plumage and behavior and is most easily identified by range. Also note Greater's larger size and slightly darker plumage.

Breeding | Adult

PIED-BILLED GREBE

Podilymbus podiceps L 13.5" (34 cm)

FIELD MARKS

Short-necked; big-headed; stocky; mostly brown plumage

Breeding adult has black ring around stout, whitish bill; black chin and throat

Winter birds lose bill ring, chin becomes white

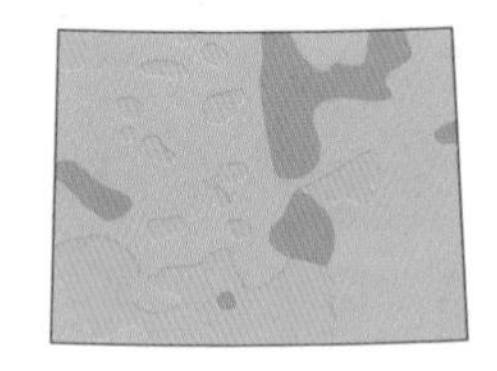

Behavior

The most secretive yet most common of North American grebes, the Pied-billed is seldom seen on land or in flight. When alarmed, it slowly sinks into the water, holding only its head above the surface. Its strong, stout bill allows it to feed on hard-shelled crustaceans, breaking apart and crushing the shells with ease. Like most grebes, it eats feathers and feeds them to its young, perhaps to protect their stomach linings from fish bones.

Habitat

Prefers nesting around lakes and ponds with marshy borders. Also along streams and rivers where cattails occur.

Local Sites

Uncommon local permanent resident at lakes and ponds throughout Colorado, becoming more widespread during migration and summer. Look for the Pied-billed Grebe wherever there are cattails, or in marsh habitats.

FIELD NOTES In winter, the Eared Grebe, *Podiceps nigricollis* (inset), is differentiated by its red eyes, its head shape, and the dark markings on its face and neck.

Breeding | Adult

WESTERN GREBE

Aechmophorus occidentalis L 25" (64 cm)

FIELD MARKS

Striking black-and-white plumage

Large; long, swan-like neck

Black cap extends below eye

Paler plumage around eye in winter

Long, pointed, yellow-green bill

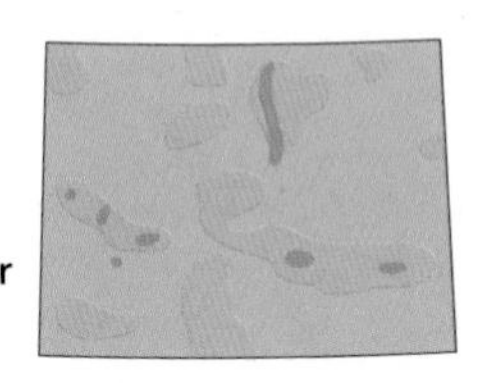

Behavior

Feeds almost exclusively on fish, which it pursues and often consumes underwater. Like herons, this grebe can snap its long neck instantaneously forward to strike or spear prey with its long, pointed bill. In courtship, a pair will rise up and rush side by side for great distances across the surface of the water. Call is a loud two-note *crick-kreek.*

Habitat

Breeds mainly inland on freshwater lakes and wetlands, where emergent vegetation borders open water. A pair cooperates in building a floating nest, anchored to the vegetation. In winter, most move to large lakes and reservoirs in Colorado.

Local Sites

View the Western Grebe at Barr Lake State Park and larger wetlands in the San Luis Valley; they are most easily seen in winter at Pueblo Reservoir.

FIELD NOTES Similar to the Western Grebe, Clark's Grebe, *Aechmophorus clarkii* (inset: breeding adult), also nests in Colorado. The Clark's Grebe has a white face that surrounds the eye, and a bright "schoolbus yellow" bill.

Breeding | Adult

AMERICAN WHITE PELICAN

Pelicanus erythrorynchos L 62" (158 cm) WS 108" (274 cm)

FIELD MARKS

Enormous size

Large orangish bill with extendible pouch

Adults all white; young birds with some brownish mottling

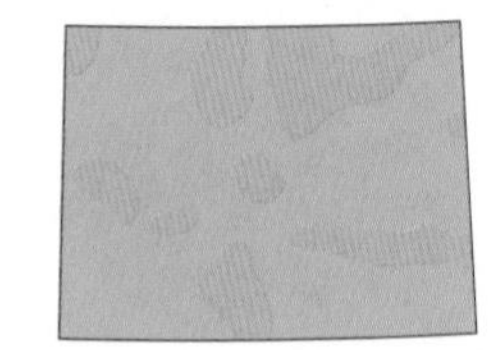

Behavior

Unlike the coastal Brown Pelican, which dives for food, American White Pelicans dip their bills into the water to scoop up fish. They frequently forage in large flocks, slowly swimming in a loose group to concentrate fish. Nesting birds gather on isolated islands in large lakes and reservoirs. Nonbreeders often congregate in large numbers at other lakes. Pelicans rarely vocalize away from the breeding ground.

Habitat

Prefers large lakes and reservoirs, but may occasionally be seen on ponds and wide, slow-moving rivers.

Local Sites

While there are fewer than five breeding sites in the state, American White Pelicans may be noted widely in migration and summer. Although they do not breed there, Barr Lake State Park is an excellent spot close to Denver to see these birds from April through September.

FIELD NOTES Numbers of American White Pelicans started rebounding in the 1960s, and this species is now seen more frequently in Colorado. The pelicans arrive in the state as early as February and individuals lingering into December have become commonplace.

Immature

DOUBLE-CRESTED CORMORANT

Phalacrocorax auritus L 32" (81 cm) WS 52" (132 cm)

FIELD MARKS

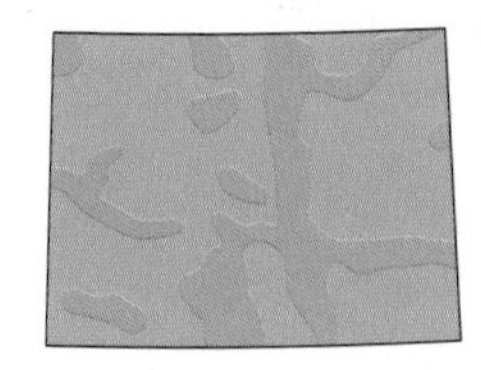

Black overall; facial skin yellow-orange; pale bill hooked at tip

Distinctive kinked neck in flight

Breeding adult has tufts of black feathers behind eyes

Immature has pale neck and breast

Behavior

After locating prey, the Double-crested Cormorant can dive to considerable depths, propelling itself with fully webbed feet. Uses its hooked bill to grasp fish. Feeds on a variety of aquatic life. When it leaves the water, it perches on a branch, dock, or piling and half-spreads its wings to dry. Soars briefly at times, its neck with a noticeable kink. May swim submerged to the neck, bill pointed slightly skyward. Emits a deep grunt.

Habitat

The most numerous and far-ranging of North American cormorants, the Double-crested may be found along coasts, inland lakes, and rivers; it adapts to fresh or saltwater environments.

Local Sites

Common, increasing breeder in Colorado, especially in the Front Range, and the South Platte and Arkansas Rivers. Rare in midwinter, when most are seen at the Pueblo Reservoir, or Valmont Reservoir in Boulder.

FIELD NOTES Double-crested Cormorants are common colonial nesters. In some locations they appear to be increasing dramatically, sometimes to the detriment of Great Blue Herons (p. 50), which may share the same breeding colonies.

Breeding | Adult

GREAT BLUE HERON

Ardea herodias L 46" (117 cm) WS 72" (183 cm)

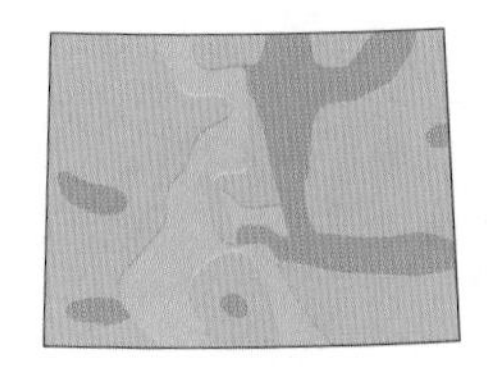

FIELD MARKS

Gray-blue overall; white foreneck with black streaks; yellowish bill

Black stripe extends above eye

Breeding adult has plumes on its head, neck, and back

Juvenile has dark crown; no plumes

Behavior

Often seen standing or wading along calm shorelines or in rivers, foraging for food. It waits for prey to come into its range, then spears it with a quick thrust of its sharp bill. Flies with its head folded back onto its shoulders in a tight S-curve, typical of other herons as well. When threatened, draws its neck back with plumes erect and points its bill at antagonist. Pairs build stick nests high in trees in loose association with other Great Blue pairs. Sometimes emits a deep, guttural squawk as it takes flight.

Habitat

May be seen hunting for aquatic creatures in marshes and swamps, or for small mammals, in fields and forest edges.

Local Sites

Found in a variety of wetland and agricultural areas throughout Colorado for most of the year.

FIELD NOTES The name "crane" is often mistakenly applied to the Great Blue Heron, but cranes belong to an altogether different family. The Sandhill Crane, *Grus canadensis* (inset), can be seen in migration across the region, particularly in the San Luis Valley. During the breeding season, cranes are found mostly in the northwest. Note its straight neck in flight.

Breeding | Adult

BLACK-CROWNED NIGHT-HERON

Nycticorax nycticorax L 25" (64 cm) WS 44" (112 cm)

FIELD MARKS

Black crown and back

White plumes on hindneck, longest when breeding

White underparts and face; gray wings, tail, and sides of neck

Immature streaked brown

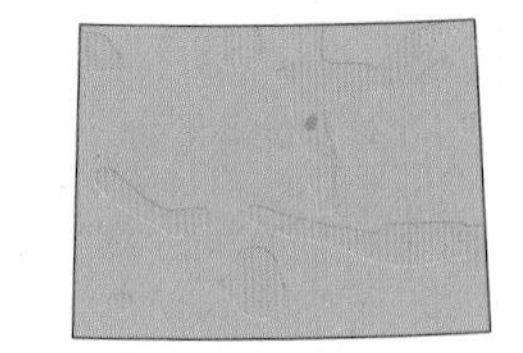

Behavior

Primarily a nocturnal feeder. Even when feeding during the day, remains in the shadows, almost motionless, waiting for prey to come within range. Forages for fish, frogs, rodents, reptiles, mollusks, eggs, and nestlings. Black-crowneds, consumers of fairly large prey, are susceptible to accumulating contaminants; their population status is an indicator of environmental quality. Call heard in flight is a gutteral *quok.*

Habitat

This heron has adapted to a wide range of habitats. In Colorado it favors freshwater wetlands and lakeshores that provide cover and forage. Nests in colonies usually in willows or other shrubs.

Local Sites

Locally common throughout much of Colorado in summer; generally scarce in winter, found regularly only along stretches of the South Platte River in the Denver area.

FIELD NOTES The juvenile Black-crowned Night-Heron (inset) is brownish overall with streaking below and pale spots above. First and second year birds become progressively more adultlike.

Breeding | Adult

WHITE-FACED IBIS

Plegadis chihi L 23" (58 cm)

FIELD MARKS

Reddish bill; red eye; red legs

White border of red facial skin extends behind eyes, under chin

Chestnut plumage with iridescent green or purplish patches on wings and crown; black wingtips

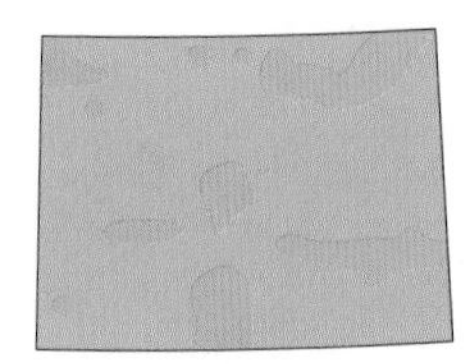

Behavior

Probes and sieves water for food with its bill. During courtship, a mating pair rubs heads together. The two offer grass and sticks to each other and preen. In flight, the White-faced Ibis appears all black. Carries neck at a downward angle, and its long legs extend well beyond its tail.

Habitat

During migration, found at lakes and reservoirs, and flooded fields in agricultural valleys. It is also commonly observed at sewage treatment ponds.

Local Sites

Common migrant wherever there is open water throughout Colorado, with large concentrations sometimes found at bigger lakes and reservoirs. The largest breeding colonies of White-faced Ibis are in the San Luis Valley.

FIELD NOTES The Glossy Ibis, *Plegadis falcinellus,* is accidental in Colorado. It looks very similar to the White-faced Ibis, especially the immature birds. The adult Glossy Ibis can be distinguished by its dark facial skin surrounded by a blue border, dark (not red) eyes, and gray-green legs.

Year-round | Adult

TURKEY VULTURE

Cathartes aura L 27" (69 cm) WS 69" (175 cm)

FIELD MARKS

In flight, two-toned underwings contrast and long tail extends beyond feet

Brownish black feathers on body; silver-gray flight feathers

Unfeathered red head; ivory bill

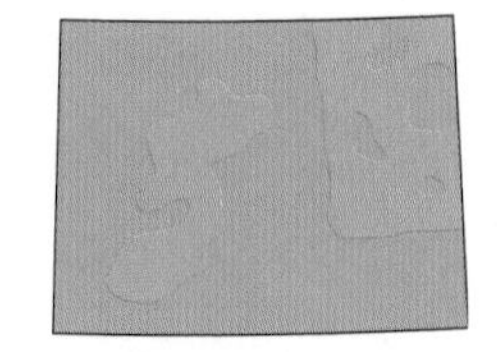

Behavior

An adept flier, the Turkey Vulture soars high above the ground in search of carrion and refuse. Rocks from side to side in flight, seldom flapping its wings, which are held upward in a shallow V, allowing it to gain lift from conditions that would deter many other raptors. Known to spread its wings wide while roosting. Well-developed sense of smell allows this species to locate carrion concealed in forest settings. Feeds heavily when food is available but can go days without. Generally silent, but will emit soft hisses and grunts while feeding.

Habitat

Hunts in open country, woodlands, farms, and also in urban dumps and landfills. Often seen over highways, searching for roadkill. Nests solitarily in abandoned buildings or hollow logs and trees.

Local Sites

Found throughout most of Colorado in summer, the Turkey vulture is most easily seen in the plateau and mesa country of western Colorado, and along the Front Range.

FIELD NOTES The Turkey Vulture's naked head is an adaptation to keep it from soiling feathers while feeding, therefore reducing the risk of picking up disease from carcasses.

Year-round | Adult, left; Third-year, right

BALD EAGLE

Haliaeetus leucocephalus L 31-37" (79-94 cm) WS 70-90" (178-229 cm)

FIELD MARKS

Distinctive white head and tail

Large yellow beak, feet, and eyes

Brown body

Juveniles mostly dark, showing blotchy white on underwing and tail

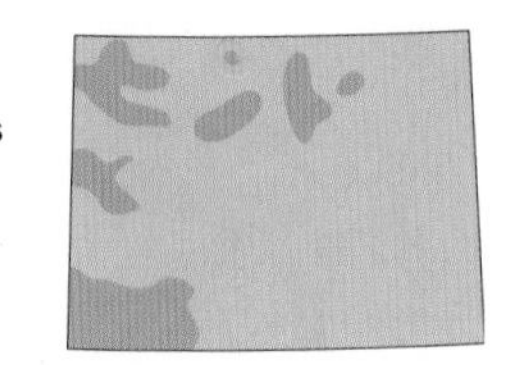

Behavior

A rock-steady flier, the Bald Eagle rarely swerves or tips on its flattened wings; it rarely even needs to flap them. Feeds mainly on fish, but sometimes on carrion or small land mammals as well. May also steal fish from other birds of prey. Bald Eagles lock talons and cartwheel together through the sky in elaborate courtship dance. Call a weak, flat, almost inaudible *kak-kak-kak*.

Habitat

This member of the sea-eagle group generally prefers reservoirs and large rivers in Colorado. It is a rare nesting species in Colorado; more common in winter. Nests solitarily in tall trees or on cliffs.

Local Sites

From late fall to early spring, look for the Bald Eagle at large water impoundments. John Martin and Jackson Reservoirs sometimes host concentrations of over 100 individuals. Denver's closest nesting pair is at Barr Lake.

FIELD NOTES The Golden Eagle, *Aquila chrysaetos* (inset), is similar to the immature Bald Eagle, but has a smaller head, a darker body, and a more sharply defined pattern on its axillaries, wing linings, and tail.

Year-round | Adult male

NORTHERN HARRIER

Circus cyaneus L 17-23" (43-58 cm) WS 38-48" (97-122 cm)

FIELD MARKS

Owl-like facial disk

White rump; long, narrow wings with rounded tips; long tail

Adult male gray above, whitish below; female brown above, whitish below, with brown streaks on breast and flanks

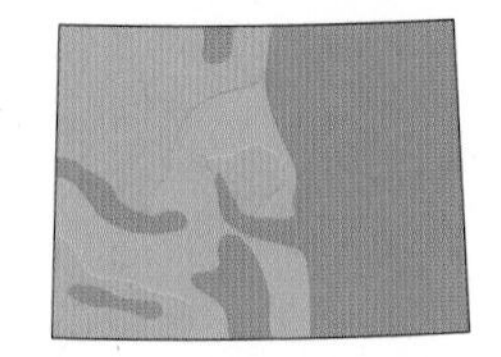

Behavior

Harriers generally perch low and fly close to the ground, wings upraised, as they search for birds, mice, frogs, and other prey. They seldom soar high except during migration and in exuberant, acrobatic courtship display, during which the male loops and somersaults in the air. Often found hunting in the dim light of dawn or dusk, using its well-developed hearing. Identifiable by a thin, insistent whistle.

Habitat

Once called the Marsh Hawk, the harrier can be found in wetlands and open country. Nests invariably on the ground. During winter months, roosts communally on the ground.

Local Sites

Colorado is home to the Northern Harrier year-round. Pawnee and Comanche National Grasslands are excellent locations to see harriers all year as they hunt over open ground.

FIELD NOTES Take care when attempting to identify a Northern Harrier high overhead. It can look like a falcon when gliding—or like an accipiter when soaring. Look for its bright white rump, one of the most noticeable field marks of any hawk.

Juvenile

SHARP-SHINNED HAWK

Accipiter striatus L 10-14" (25-36 cm) WS 20-28" (51-71 cm)

FIELD MARKS

Adult blue-gray above, reddish brown streaks on neck, breast, and belly

Squared-off tail with narrow white tip

Thin, bright yellow legs and feet

Juveniles are brown above, white below with brown streaking

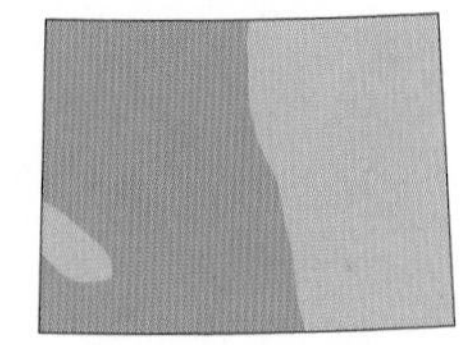

Behavior

Preys chiefly on small birds, often engaging in ambush maneuvers or aggressive pursuit even through thick foliage and undergrowth. Flight consists of several quick wing beats and a glide, its quick turns assisted by a long, rudderlike tail. The small hawk is highly aggressive, even against humans, when defending its territory. An alarmed bird calls *kek-kek-kek.*

Habitat

Found in mixed woodlands, but can also be seen in the open, especially during migration. Nests are substantial stick structures located in tall trees.

Local Sites

The Sharp-shinned can be found in suburban areas throughout Colorado in winter. During summer, the species retires to coniferous and mixed coniferous forests.

FIELD NOTES Distinguishing the Sharp-shinned from the Cooper's Hawk, *Accipiter cooperii* (inset: juvenile, top; adult, bottom), is one of birding's classic challenges. Both species are largely brown as juveniles; blue-gray above, rufous below as adults. The Cooper's is slightly larger, has a more rounded tail, and appears larger headed in flight.

Year-round | Adult light morph

SWAINSON'S HAWK

Buteo swainsoni L 21" (53 cm), WS 52" (132 cm)

FIELD MARKS

Long, narrow pointed wings

Light morph: dark bib, whitish wing linings, dark flight feathers

Dark morph: dark body, white undertail coverts, dark wings

Behavior

The Swainson's Hawk soars over open plains and prairies with teetering, vulture-like flight. Its primary prey consists of large insects and small rodents, such as mice and ground squirrels. Call is a high-pitched, raspy *kreeee.*

Habitat

Frequents open country, such as grassland, scrubland and agricultural fields. Nest is a mass of sticks lined with leafy twigs, grass, weeds, and wool, placed in a tree or on the ground.

Local Sites

A common migrant and summer resident in open grasslands and agricultural fields in eastern Colorado and the mountain parks. Generally less common on the West Slope; most frequently seen there in migration. There are no winter records of Swainson's Hawk in Colorado.

FIELD NOTES Most Swainson's Hawks—including those that nest in Canada—migrate to grasslands in Argentina for the winter. Huge flocks are seen along their southbound route where the landmass narrows, such as Veracruz, Mexico. Single day counts of over 300,000 birds have been tallied.

Year-round | Adult light morph

RED-TAILED HAWK

Buteo jamaicensis L 22" (56 cm) WS 50" (127 cm)

FIELD MARKS

Brown above; red tail on adults

Whitish belly with broad band of dark streaking

Dark brown leading edge of underwing

Immature has brown, banded tail

Behavior

Watch the Red-tailed Hawk circling above, searching for rodents, sometimes kiting, or hanging motionless on the wind. Uses thermals to gain lift and limit its energy expenditure while soaring. Perches for long intervals on telephone poles and other man-made structures, often in urban areas. Listen for its distinctive call, a harsh, descending *keee-eeer.*

Habitat

Seen in more habitats than any other North American buteo, from woods to prairies to deserts. Common at habitat edges, where field meets forest or wetlands meet woodlands, favored for the variety of prey found there.

Local Sites

Common throughout the state. Scan the edges of open areas for a perched Red-tailed surveying its territory.

FIELD NOTES In winter, Colorado's Red-tailed Hawks are augmented with birds from the north. Some of these are dark morph, rufous morph, or intermediate individuals. The *harlani* subspecies (inset) is usually very dark with a more grayish tail and white mottling on its breast.

Year-round | Adult light morph

FERRUGINOUS HAWK

Buteo regalis L 23" (58 cm) WS 53" (135 cm)

FIELD MARKS

Large size with yellow gape

Adults have rufous-colored legs, juveniles have white legs.

In flight, note long, fairly broad, pointed wings, angled slightly upward

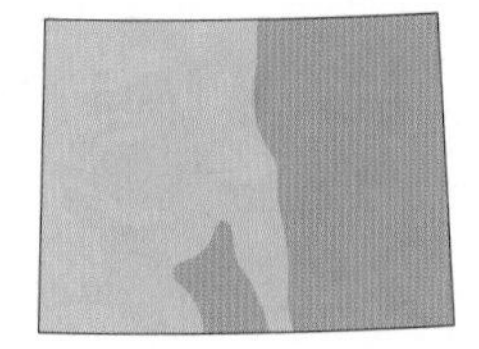

Behavior

Because Ferruginous Hawks are found in open areas, nesting sites are severely limited. Most individuals build large stick nests in an isolated tree. Nests are often constructed of sticks and stems from sagebrush and other large shrubs. The male gives a series of long drawn-out whistles on the breeding grounds that is reminiscent of a Lark Bunting. The calls heard year round are a low harsh *krrr.*

Habitat

The Ferruginous Hawk is a bird of the open country, preferring grasslands and badlands. Prairie dog towns are favored sites. They frequently perch on utility poles, but are also regularly seen perched on the ground.

Local Sites

Found in open grassland areas mostly in eastern Colorado. The Pawnee National Grasslands is an excellent year-round location for observing Ferruginous Hawks. While not their preferred habitat, Ferruginous Hawks may also be seen at open areas near cities, usually in the winter.

FIELD NOTES Ferruginous Hawks are more frequently seen from October to March, when local breeding birds are supplemented by migrants and wintering individuals. During winter and migration a small percentage of birds are dark or rufous-morph birds.

Year-round | Adult male

AMERICAN KESTREL

Falco sparverius L 10.5" (27 cm) WS 23" (58 cm)

FIELD MARKS

Russet back and tail; streaks or dots on pale underparts

Two black stripes on white face

Male has blue-gray wing coverts

Female has russet wing coverts and russet streaks on her breast

Behavior

Feeds on insects, reptiles, and mice and other small mammals. Can remain stationary over prey by coordinating its flight speed with the wind speed, then plunges down for the kill. Will also feed on small birds, especially in winter. Regularly seen perched on fences and telephone lines, bobbing its tail with frequency. Has clear, shrill call of *killy-killy-killy.*

Habitat

North America's most widely distributed falcon. Found in open country and in cities, often "mousing" along highway medians or guarding small pastures. Nests in tree holes, barns, or man-made boxes using little or no nesting material.

Local Sites

The most common and widespread falcon in the region, American Kestrels can be found in most areas of Colorado throughout the year.

FIELD NOTES The Prairie Falcon, *Falco mexicanus* (inset: adult), an uncommon and local breeding bird in much of Colorado, winters in agricultural areas where American Kestrels are found. The Prairie Falcon is larger and brownish overall, with dark axilliaries and underwing coverts.

Year-round | Adult

AMERICAN COOT

Fulica americana L 15.5" (39 cm)

FIELD MARKS

Blackish head and neck; slate gray body

Small, reddish brown forehead shield

Whitish bill with dark band at tip

Greenish legs with lobed toes

Behavior

The distinctive toes of the American Coot are flexible and lobed, permitting it to swim well in open water and even to dive in pursuit of aquatic vegetation and invertebrates. It has the ability to tip its tail up and stay submerged to feed. Bobs its small head back and forth when walking or swimming; forages in large flocks, especially during the winter. Has a wide vocabulary of grunts, quacks, and chatter.

Habitat

Nests in freshwater marshes, in wetlands, or near lakes and ponds. Less common in winter. The American Coot has also adapted well to human-altered habitats, including sewage lagoons for foraging and suburban lawns for roosting.

Local Sites

Easily found at wetlands throughout Colorado, such as Barr Lake State Park and Pueblo Reservoir.

FIELD NOTES Its body too heavy for direct takeoff, the American Coot's lobed toes help it to "run" on water. Accelerating with its wings flapping rapidly, it is able to gain the speed it needs to take flight.

Year-round | Adult

KILLDEER

Charadrius vociferus L 10.5" (27 cm)

FIELD MARKS

Gray-brown above; white neck and belly; two black breast bands

Black stripe on forehead and one extending back from black bill

Red-orange rump visible in flight

Red orbital ring

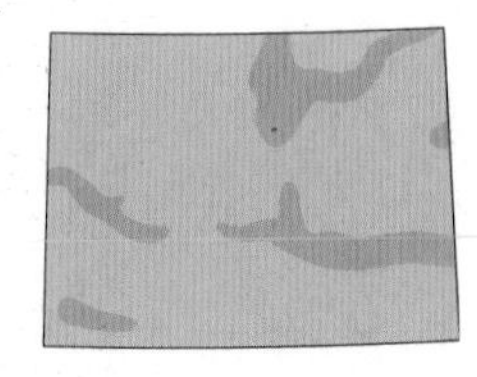

Behavior

Often seen running, then stopping abruptly with an inquisitive look, then jabbing at the ground with its bill. Feeds mainly on insects. May gather in loose flocks. Its loud, piercing, eponymous call of *kill-dee,* or its rising *dee-dee-dee,* is often the signal for identifying these birds before seeing them. Listen also for a long, trilled *trrrrrrr* during courtship displays or when a nest is threatened by a predator.

Habitat

Although a type of plover—one of the shorebirds—the Killdeer is most often found in grassy regions; it builds its nest on almost any patch of open ground.

Local Sites

One of Colorado's most ubiquitous shorebirds, the Killdeer is found in open areas across the state from spring through fall and winters along rivers and in open wetlands.

FIELD NOTES If its nest is threatened by an intruder, the Killdeer is known to feign a broken wing, limping to one side, dragging its wing, and spreading its tail in an attempt to lure the threat away from its young. Once the predator is far enough away from the nest, the instantly "healed" Killdeer takes flight.

Breeding | Adult

MOUNTAIN PLOVER

Charadrius montanus L 20" (51 cm)

FIELD MARKS

Brown upperparts; plain whitish underparts; buff tinge to breast

In flight: white underwings; dark tip to tail

Breeding: black forecrown, eyeline

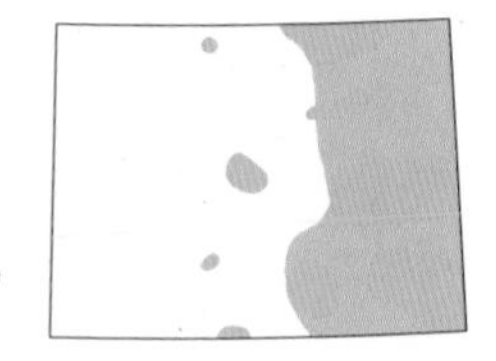

Behavior

Usually found singly or in pairs. Its nest is a simple scrape on the ground, excavated by the male. In late summer migrating flocks numbering into the hundreds are occasionally seen along dry lakeshores and playas. On the breeding grounds Mountain Plovers give a series of low, prolonged whistles that are somewhat reminiscent of a singing Lark Bunting (p. 217). The call note heard year-round is a low rolling *krrr.*

Habitat

A classic symbol of the shortgrass prairie. The Mountain Plover favors the very short grass created by prairie dogs, bison, cattle, and other herbivores.

Local Sites

Mountain Plovers arrive in Colorado in early March, with most departing the state by late August. Look for breeding birds at Pawnee National Grasslands, particularly where prairie dogs abound.

FIELD NOTES The name Mountain Plover is something of a misnomer, as the species is rarely seen in the mountains away from the mountain parks. Within the "mountains," South Park has the largest numbers of Mountain Plovers, but here too the species is found in open areas with very short grass. The Mountain Plover is thought to have declined greatly in the last century due to loss of habitat.

Breeding | Adult male

AMERICAN AVOCET

Recurvirostra americana L 18" (46 cm)

FIELD MARKS

Rusty head and neck become grayish in winter

Black and white back; white belly; long, bluish legs and feet

Sharply upcurved bill, longer and straighter in males

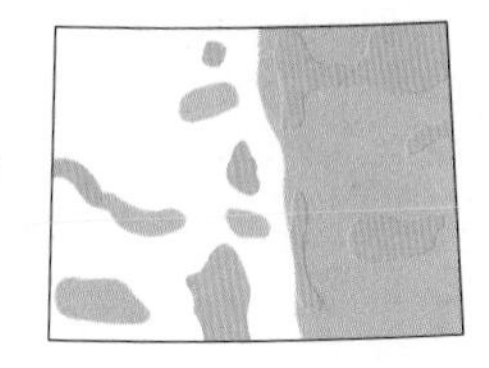

Behavior

This graceful wader feeds with a flock in shallow water by walking in a loose line of sometimes 100 birds, sweeping its slightly open bill in a scything motion just below the water's surface, filtering out small organisms such as aquatic larvae and small crustaceans. In deeper water, may feed by tipping up, much like a dabbling duck. Its bill is so sensitive that it will defend itself solely with wings and feet. Call is a loud *wheet* or *pleeet.*

Habitat

Prefers shallow alkaline lakes and briny ponds for foraging. Nests on flat ground near water.

Local Sites

Avocets arrive in late March and remain through October, with numbers peaking in late April and July. They can be seen at wetlands in eastern Colorado and the mountain parks.

FIELD NOTES The only other North American member of the family Recurvirostridae, the Black-necked Stilt, *Himantopus mexicanus* (inset), has exceptionally long, pinkish red legs, which enable it to forage with its needle-thin bill in waters deeper than most shorebirds. Inhabiting a more limited range in Colorado than the avocet, the stilt is easily distinguished by its all-black back and its white cheeks.

Breeding | Adult

SPOTTED SANDPIPER

Actitis macularia L 7½" (19 cm)

FIELD MARKS

Brown upperparts, barred in breeding plumage

White underparts, spotted brown in breeding plumage

Tail shorter than other sandpipers

Short white wing stripe

Behavior

The somewhat larger female is the first to establish territory and defend it during breeding season. She may also mate with several males in a season. Males will tend to the eggs and young. Feeds on insects or crustaceans and other invertebrates by plucking them from the water's surface or even snatching them from the air. Walks with a nodding or teetering motion.

Habitat

One of the most common and widespread sandpipers in North America during breeding season, preferring sheltered ponds, lakes, streams, or marshes.

Local Sites

Most common in summer along mountain streams and wetlands, including areas above timberline. Easily seen in migration throughout Colorado.

FIELD NOTES During winter, the Spotted Sandpiper (inset: winter) lacks its distinctive spotting. It can still be easily identified by its characteristic tail-bobbing behavior, dark eye line, and white supercilium.

Juvenile

BAIRD'S SANDPIPER

Calidris bairdii L 7.5" (19 cm)

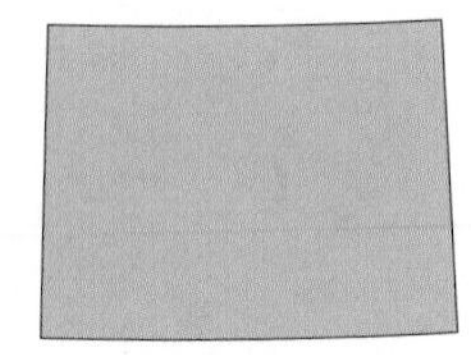

FIELD MARKS

Small size; overall brownish upperparts; pale underparts

Brownish streaks across breast

Long wings project past the tip of the tail; black legs

Behavior

Forages for small prey with its short, thin bill in muddy, sandy, or shallow water. Not particularly wary. Call is a short burry *jeuurk.*

Habitat

Baird's Sandpiper can be found at just about any water body with a shore in Colorado, from lakes and ponds to sewage treatment plants—infrequently even along wide rivers with mudflats. Aside from the Spotted Sandpiper that breeds along alpine streams, this is the most likely shorebird to be seen on alpine tundra. Also occasionally found on dry pastures and golf courses. Breeds in the Arctic.

Local Sites

Perhaps the state's most abundant migrant shorebird, Baird's Sandpipers appear chiefly from late March through May, and again from July through early October. During these times they are most easily observed at reservoirs with abundant mudflats in eastern Colorado.

FIELD NOTES The Least Sandpiper, *Calidris minutilla* (inset: juvenile), is similar to Baird's but smaller, with yellowish to greenish legs, and shorter wings that do not project past the tip of the tail.

Year-round | Adult

WILSON'S SNIPE

Gallinago delicata L 10.25" (26 cm)

FIELD MARKS

Stocky with very long bill; very short tail; pointed wings

Head and back boldly striped blackish brown and pale buff

Heavily barred flanks

Dark underwings, white belly

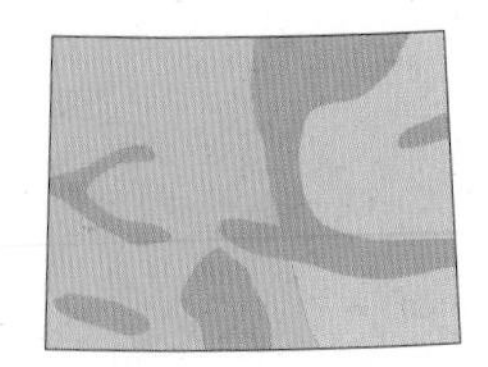

Behavior

Often not seen until flushed, when it gives harsh *ski-ape* call and rapidly flies off in a zigzag pattern. Feeds on insects, larvae, and earthworms by probing mud with its bill in a jerky fashion. Generally solitary, the snipe does not interact with other shorebirds. In swooping flight display, known as "winnowing," a quavering hoot-like sound is produced by air vibrating the two outermost tail feathers.

Habitat

Found in freshwater marshes and swamps and in any damp, muddy wetland where cover is afforded by vegetation. May frequent open areas as well. Nests on the ground in a scrape.

Local Sites

Breeds in damp areas in the mountains and plains. The few snipe that winter over in Colorado usually remain at lower elevations near open waterways.

FIELD NOTES With a bill that can rival that of the Wilson's Snipe in length and straightness, the Long-billed Dowitcher, *Limnodromus scolopaceus* (inset: nonbreeding), isdistinguished by its gray winter plumage, its V-shaped white patch extending up its back, and its tendency to forage in flocks.

Breeding | Adult female

WILSON'S PHALAROPE

Phalaropus tricolor L 9.25" (24 cm)

FIELD MARKS

Long, black needlelike bill

In breeding plumage, females more colorful than males; both with blackish stripe on face and neck

White rump visible in flight

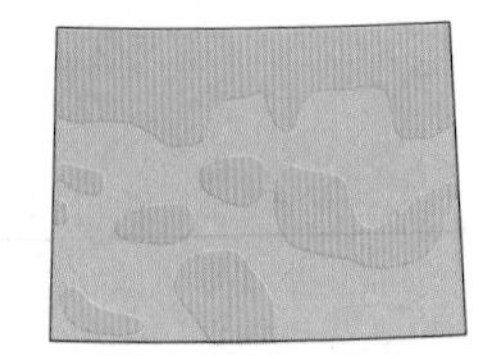

Behavior

Feeds on land and water. Foraging while swimming, this species sometimes whirls in a circle to create a vortex that brings small prey to the surface. Also probes through mud and captures flying insects. Females, more aggressive than the males, compete in courtship displays, stretching out their necks and puffing their feathers. Females migrate first, leaving the males on the breeding grounds to incubate eggs and care for the young.

Habitat

Migrants found wherever there is open water with sandy or muddy borders.

Local Sites

Common to abundant migrant at lakes, reservoirs, and sewage treatment plants throughout Colorado, from April through September. Plains reservoirs such as Jackson, John Martin, and Prewitt often harbor concentrations of hundreds, even thousands, of phalaropes in late summer.

FIELD NOTES Winter plumage (inset), often seen during fall migration, is gray above with dark primaries and a white rump. In this plumage males and females are similar, although the females are larger.

Breeding | Adult

FRANKLIN'S GULL

Larus pipixcan L 14-15" (36–38 cm) WS 35-38" (89–97 cm)

FIELD MARKS

Breeding adult: black hood; reddish bill; white underparts; slaty gray wings; black primaries with white patterning

Winter adult: partial hood; blackish bill with orange tip

Behavior

Forages for a variety of foods from invertebrates to decaying fish to garbage at the local dump. Franklin's Gulls are often found in large flocks. Its short rounded wings help it make quick aerial turns that aid in foraging and allow it to quickly drop down to nesting sites located in marshes.

Habitat

Commonly seen from April to October at lakes and reservoirs throughout Colorado. Also regularly found foraging in agricultural areas. Winters in South America.

Local Sites

This species is easily seen at large reservoir, wetlands and even behind combines working in agricultural areas. While very few breed in the state, some nonbreeding birds remain all summer, and by July through early October the species is often abundant at Prewitt Reservoir and other large water bodies on the eastern plains.

FIELD NOTES The Bonaparte's Gull, *Larus philadelphia* (inset: breeding adult), is similar, but smaller with a pale gray mantle and white outer primaries. In Colorado, Bonaparte's Gulls are rarely seen away from reservoirs and larger bodies of water.

Nonbreeding | Adult

RING-BILLED GULL

Larus delawarensis L 17.5" (45 cm) WS 48" (122 cm)

FIELD MARKS

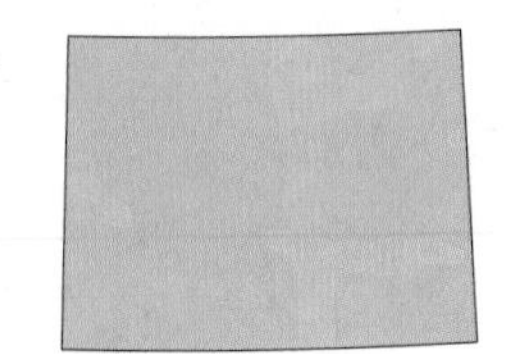

Adult: Yellow bill with black subterminal ring; pale eye with dark orbital ring

Pale gray upperparts; white underparts; yellowish legs; black primaries show white spots

Head streaked brown in winter

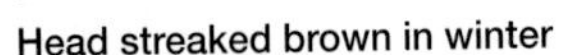

Behavior

This opportunistic feeder will scavenge for garbage, grain, dead fish, fruit, and aquatic invertebrates, often demanding scraps of food from picnickers. A vocal gull, it calls, croaks, and cries incessantly, especially while feeding. Its call consists of a series of laughing croaks that begins with a short, gruff note and falls into a series of *kheeyaahhh* sounds. The Ring-billed takes three years to attain full adult plumage; the first winter Ring-billed has a pinkish bill with a dark tip.

Habitat

A regular visitor to most bodies of water, especially reservoirs in urban areas, but equally at home at dumps and fast-food parking lots.

Local Sites

Easily seen, particularly along the Front Range. Dumps and reservoirs are all favorites of the Ring-billed Gull.

FIELD NOTES The Herring Gull, *Larus argentatus* (inset), is similar to the Ring-billed Gull but larger, with pink legs, and red and black on its bill. Juveniles are entirely brown and gradually acquire adult plumage over four years.

Nonbreeding | Adult

CALIFORNIA GULL

Larus californicus L 21" (53 cm) WS 54" (137 cm)

FIELD MARKS

Adult: white head, heavily streaked with brown in winter; white underparts

White neck and breast; gray back;yellow bill with black and red spots on lower mandible

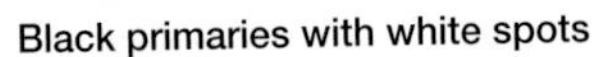

Black primaries with white spots

Behavior

Forages on insects, rodents, eggs and chicks, refuse, and carrion. This species of gull saved the Mormon settlement in Utah by consuming the hordes of grasshoppers that swarmed their agricultural fields. The California Gull is the Utah state bird. Its long call is a series of *kyow* notes; the first two longer and more drawn out. Call is higher pitched than that of the Herring Gull.

Habitat

Found at reservoirs and landfills in Colorado during the summer; most leave the state in winter.

Local Sites

Look for this gull year-round along the Front Range, although numbers decline greatly in midwinter when the species is most regularly seen at Pueblo Reservoir.

FIELD NOTES The California Gull takes four years to acquire adult plumages. Once the species has acquired a gray mantle, it is darker mantled than Ring-billeds or Herring Gulls (both p. 91). Also note the darker eye and greenish yellow legs of adult California Gulls.

Year-round | Adult

ROCK PIGEON

Columba livia L 12.5" (32 cm)

FIELD MARKS

Variably plumaged, with head and neck usually darker than back

White cere at base of dark bill, pink legs

Iridescent feathers on neck reflect green, bronze, and purple

Behavior

Feeds during the day on grain, seeds, fruit, and refuse in cities, suburbs, parks, and fields; a frequent visitor to farms and backyard feeding stations as well. As it forages, the Rock Pigeon moves with a short-stepped, "pigeon-toed" gait while its head bobs back and forth. Courtship display consists of the male puffing out neck feathers, fanning his tail, and turning in circles while cooing; results in a pairing that could last for life. Characterized by soft *coo-cuk-cuk-cuk-cooo* call.

Habitat

Anywhere near human habitation. Nests and roosts primarily on high window ledges, on bridges, and in barns. Builds nest of stiff twigs, sticks, and leaves.

Local Sites

Introduced from Europe by settlers in the 1600s, the Rock Pigeon is now widespread and abundant throughout most developed regions of North America.

FIELD NOTES The Rock Pigeon's variable colors range from rust red to all white to mosaic (inset) due to centuries of domestication. Those resembling their wild ancestors have a dark head and neck, two black wing bars, a white rump, and a black terminal band on the tail.

Year-round | Adult

BAND-TAILED PIGEON

Patagioenas fasciata L 14.5" (37 cm)

FIELD MARKS

Purplish head and breast

Dark-tipped yellow bill; yellow legs

Broad gray terminal tail band

Narrow white band on nape of adult, bordered below by a spot of greenish iridescence

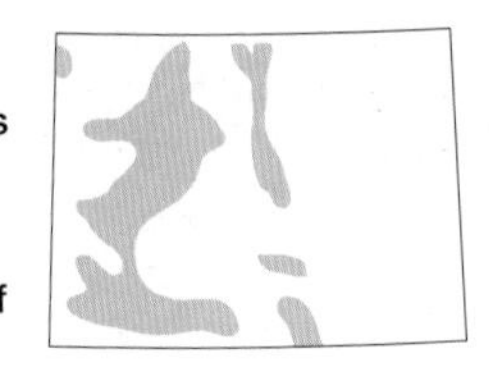

Behavior

Often perches for long spans of time either singly or in a small group at the tops of trees with little or no foliage. Size of flock may increase in winter. Forages among branches for berries, grains, seeds, nuts, and insects; rarely descends to the ground. Call is a low, repetitive *whoo-whoo,* that sounds vaguely owl-like. During breeding season, male calls from an open perch to attract a mate.

Habitat

Found primarily in forests of tall coniferous trees located in montane areas. The Band-tailed is becoming increasingly common in suburban parks. Nests on platform of twigs in the fork of a tree.

Local Sites

Most commonly found in the southwestern part of the state, but regularly occurs along the Front Range.

FIELD NOTES The Band-tailed Pigeon was nearly hunted to extinction in the 20th century until restrictive measures were implemented in the 1980s. The species has recovered, but is still a cause of concern for conservationists.

Year-round | Adult

MOURNING DOVE

Zenaida macroura L 12" (31 cm)

FIELD MARKS

Gray-brown; black spots on upper wings; white on outer tail feathers shows in flight

Trim-bodied; long pointed tail

Black spot on lower cheek; pinkish wash on neck

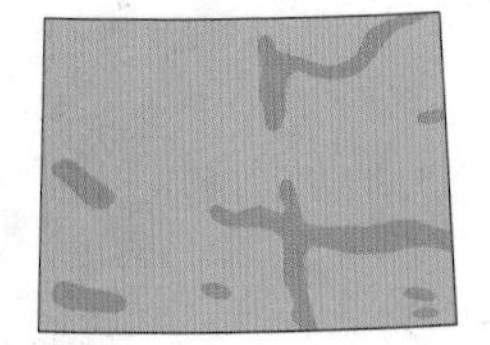

Behavior

Generally a ground feeder, the Mourning Dove forages for grain, seeds, grasses, and insects. Like other Columbidae, it is able to siphon up water without tipping back its head. Also able to produce "pigeon milk" in its crop lining, which it regurgitates to young during their first few days. Wings produce fluttering whistle as the bird takes flight. Known for mournful call, *oowooo-woo-woo-woo,* given by males in breeding season.

Habitat

Widespread and abundant, the Mourning Dove is found in a variety of habitats, but prefers open areas, often choosing suburban sites for feeding and nesting.

Local Sites

Abundant throughout Colorado, from wooded settings to farm fields to cities and towns. Most doves leave for the winter.

FIELD NOTES An introduced species, the Eurasian Collared-Dove, *Streptopelia decaocto* (inset), has recently spread into Colorado. Its tail is more rounded than the Mourning's when perched, and its primaries are darker in flight. Look as well for its black collar edged in white.

Year-round | Adult

GREAT HORNED OWL

Bubo virginianus L 22" (56 cm)

FIELD MARKS

Large; mottled brownish gray above; densely barred below

Long ear tufts (or "horns")

Rust-colored facial disk

Yellow eyes; white chin and throat; buff-colored underwings

Behavior

Chiefly nocturnal. Feeds on a variety of animals including rodents, skunks, porcupines, birds, snakes, grouse, and frogs; watches from high perch, then swoops down on prey. One of the earliest birds to nest, beginning in January or February, possibly to take advantage of winter-stressed prey. Call is a series of three to eight loud, deep hoots, the second and third often short and rapid.

Habitat

The most widespread owl in North America, the Great Horned Owl can be found in a wide variety of habitats including forests, cities, and farmlands. Reuses abandoned nests of other large birds.

Local Sites

Fairly common and widespread resident throughout Colorado in a variety of forested habitats at virtually all elevations.

FIELD NOTES The Long-eared Owl, *Asio otus* (inset), a rare and local breeding bird and uncommon winter resident, is smaller than the Great Horned Owl and has longer, more closely set ear tufts. Its rarely heard vocalization is one or more long hoots.

Year-round | Adult

BURROWING OWL

Athene cunicularia L 9.5" (24 cm)

FIELD MARKS

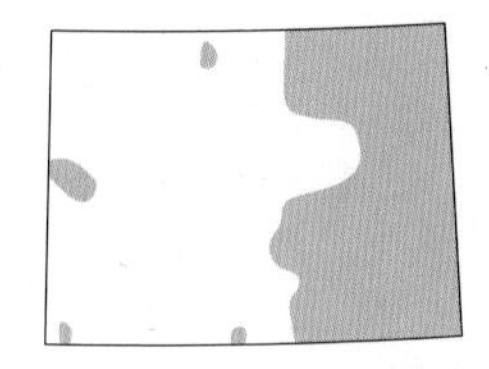

Brown above with white spotting on back, white streaks on crown

White below with brown bars

Large, yellow eyes topped by broad white eyebrows

Long legs

Behavior

A terrestrial owl, forages primarily at night, dawn, and dusk on insects and small mammals such as mice. Flight is low and undulating; can hover like a kestrel. Regularly perches on the ground or on low posts during the day near its burrow. Nests in single pairs or small colonies often in prairie dog towns, enlarging and reshaping the small mammals' already existing burrows by kicking dirt out backwards. Calls include a soft *co-coooo* and a chattering series of *chack* notes.

Habitat

Open grasslands and prairies. Nests in abandoned burrows of prairie dogs, ground squirrels, or gophers.

Local Sites

Though declining throughout its North American range, the Burrowing Owl can still be found breeding in eastern Colorado. Now extremely local on the West Slope. Look for it at Pawnee and Comanche National Grasslands and around Barr Lake.

FIELD NOTES If disturbed at the nest, both parent and juvenile Burrowing Owls often give alarm call of a rapid chatter, which imitates the sound of a rattlesnake's rattle.

Year-round | Adult

COMMON NIGHTHAWK

Chordeiles minor L 9.5" (24 cm)

FIELD MARKS

Dark gray-brown mottled back; bold white bar across primaries

Long, pointed wings with pale spotting; tail slightly forked

Underparts whitish with bold dusky bars; bar on tail in males

Behavior

The Common Nighthawk's streamlined body allows agile aerial displays when feeding at dusk. Hunts in flight, snaring insects. Drops lower jaw to create opening wide enough to scoop up large moths. Skims over surface of lakes to drink. Roosts on the ground, scraping a shallow depression, or on branches, posts, or roofs. Call is a nasal *peent.* Male's wings make hollow booming sound during diving courtship display.

Habitat

Frequents woodlands and shrubby areas; also seen in urban and suburban settings. Nests on the ground or on gravel rooftops.

Local Sites

Common but declining summer visitors to Colorado. Listen for their froglike calls in open habitats. They are easily seen in summer at Pawnee National Grasslands, or perching in trees at Crow Valley Campground.

FIELD NOTES The Common Nighthawk only became common in towns and cities of North America in the mid-1800s with the development of flat, graveled roofs, upon which it could nest. Today it is seen regularly in summer swooping past street lights, snagging flying insects drawn to the outdoor luminescence.

Year-round | Adult

WHITE-THROATED SWIFT

Aeronautes saxatalis L 6.5" (17 cm)

FIELD MARKS

White chin, breast, and center belly

White oval patch on flanks

Black upperparts

Long, forked tail; looks pointed when not spread

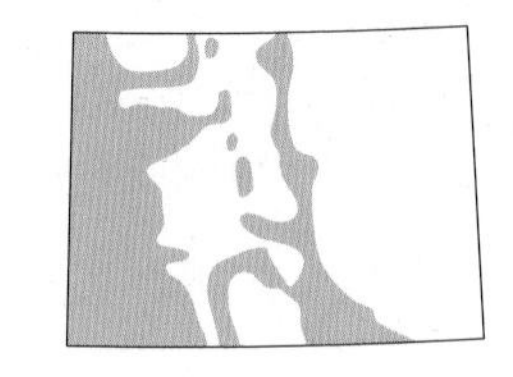

Behavior

Fast-flying bird often seen zooming around cliffs and mesas in small noisy flocks. Because of legs and feet poorly adapted to the ground, the swift does not perch during the day. It spends the time in flight, foraging for insects and ballooning spiders. Nests in crevices in cliffs, rock walls, and even tall urban buildings and highway overpasses. Ranges widely during the day, traveling many miles on feeding forays. Call is a shrill chatter, *ji-ji-ji-jijiji.*

Habitat

Common near mountains, canyons, and cliffs. Because they fly so fast and forage so widely, they are also seen over open areas and flatlands far from their nest sites.

Local Sites

Common year-round throughout the mesas and plateaus of western Colorado; often seen foraging near ponds and lakes, particularly during spring storms. Look for it at Red Rocks Park and other locations along the Front Range.

FIELD NOTES The White-throated Swift is the most widespread breeding swift in Colorado. Listen for its high twittering vocalizations wherever there are high cliffs.

Year-round | Adult male

BROAD-TAILED HUMMINGBIRD

Selasphorus platycercus L 4" (10 cm)

FIELD MARKS

Male has green crown and rose-red gorget; mostly black tail

Female white throat with speckling; buff washed underparts; a little rufous at base of tail

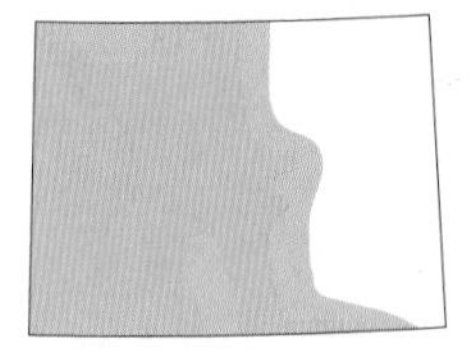

Behavior

Forages in flowers and foliage for nectar, small insects, and spiders. Males defend breeding territory. Usually nests on low horizontal tree branch, often above water. Female lays eggs in June or July, incubates them for 16 to 17 days, and tends her young alone. Chicks fledge in 21 to 26 days. Except during winter molt, male's wingbeats produce a loud, metallic trill, resembling a cricket. Call is a metallic *chip*.

Habitat

Open woodlands, especially those with pinyons, junipers, conifers, and aspens; brushy hillsides and montane scrub and thickets.

Local Sites

Abundant spring and summer resident at higher elevations in all of the major mountain ranges. Becomes more widespread at lower elevations during fall migration, but still rare on Colorado's far eastern plains.

FIELD NOTES The female Calliope Hummingbird, *Stellula calliope* (inset: female, left; male, right), resembles the female Broad-tailed in her green upperparts and pale buff underparts, but has less rufous in her tail and her wings are as long as her tail. The male Calliope differs from the Broad-tailed male in his streaked gorget.

Year-round | Adult male

RUFOUS HUMMINGBIRD

Selasphorus rufus L 3.75" (10 cm)

FIELD MARKS

Male rufous above, rufous wash below; dark orange-red gorget

Female has green upperparts and white, speckled throat

Both show white breast patch extending down center of belly

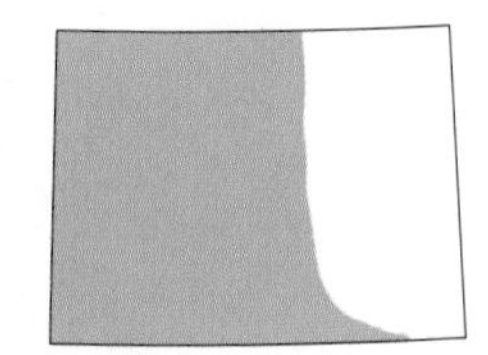

Behavior

Makes one of the longest migrations relative to body size of any bird, needing frequent refueling. This highly aggressive hummer drives away competitors up to three times its size, including blackbirds, thrushes, even chipmunks. Courtship flight consists of male ascending with back to female, then diving and turning on its whistling wings with iridescent gorget showing. Calls include a *chip*, often given in a series, and a chase note of *zeee-chuppity-chup.*

Habitat

Found from late June through mid-September, mostly in forested areas in the mountains and the West Slope.

Local Sites

Rufous and Calliope Hummingbirds are drawn to backyard feeders. During drought years, they can even be found at feeders in the Denver metropolitan area.

FIELD NOTES The female Rufous Hummingbird (inset), is every bit as feisty as the male. Note the rich rufous on her flanks, which contrasts boldly with her green upperparts.

Immature | Male

BELTED KINGFISHER

Ceryle alcyon L 13" (33 cm)

FIELD MARKS

Blue-gray head with large, shaggy crest

Blue-gray upperparts and breast band; white underparts and collar

Long, heavy bill

Chestnut sides and belly band in female

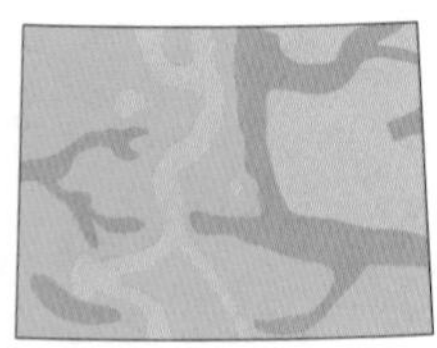

Behavior

Generally solitary and vocal, dives headfirst for fish from a waterside perch or after hovering above in order to line up on its target. Also feeds on insects, amphibians, and small reptiles. Both male and female Belteds carry out work in excavating nest tunnel, and share parenting duties. Mated pairs renew their relationship each breeding season with courtship rituals such as dramatic display flights, the male's feeding of the female, and vocalizations. Call is a loud, dry rattle; it is given when alarmed, to announce territory, or while in flight.

Habitat

Conspicuous along rivers, ponds, and lakes. Prefers partially wooded areas. Monogamous pairs nest in burrows they dig three or more feet into vertical earthen banks near watery habitats.

Local Sites

Found year-round throughout most of Colorado in riparian areas such as Chatfield State Recreation Area, Barr Lake State Park, and Wheatridge Park.

FIELD NOTES The Belted Kingfisher female is one of the few birds in North America that is more colorful than her male counterpart, which lacks the female's chestnut band across the belly and chestnut sides and flanks.

Year-round | Adult

RED-NAPED SAPSUCKER

Sphyrapicus nuchallis L 8.5" (22 cm)

FIELD MARKS

White markings on back in two rows

Usually red patch on nape

Red crown and throat

Juveniles similar pattern to adult, but without bright coloration.

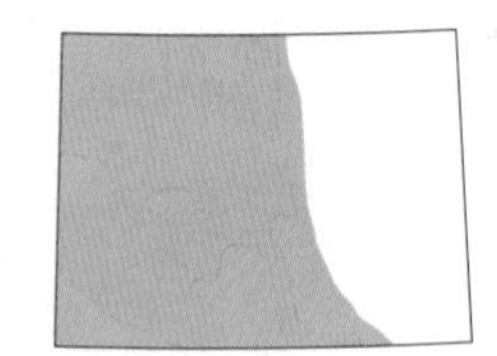

Behavior

Sapsuckers make a series of small holes in trees, called wells, using long tongues to sip the sap, aided by the small hairlike projections on the tip. These sap wells attract insects, which in turn attract other birds, particularly hummingbirds. Sapsuckers have a distinctive drumming pattern: a burst of rapid taps followed by gradually slowing taps, interspersed at the end with double taps. Their call is loud nasal *queeh.*

Habitat

In Colorado, the Red-naped is most often found in aspen or mixed woodlands, or along mountain streams and rivers. It will often create sap wells in willows. In migration, it is found in a wider variety of habitats.

Local Sites

Listen for its nasal calls or drumming at Endovalley Picnic Area in Rocky Mountain N.P. Commonly seen in or near groves of aspen and other deciduous trees.

FIELD NOTES Colorado's other breeding sapsucker, the Williamson's, *Sphyrapicus thyroidus* (inset: male, left; female, right), is usually found in ponderosa pines and other coniferous trees. Like the Red-naped Sapsucker, Williamson's departs the state in winter.

Year-round | Adult female

DOWNY WOODPECKER

Picoides pubescens L 6.75" (17 cm)

FIELD MARKS

Black cap, ear patch, malar stripe; black wings with white spotting

Patch of white on back

White tuft in front of eyes; white underparts

Red occipital patch on male only

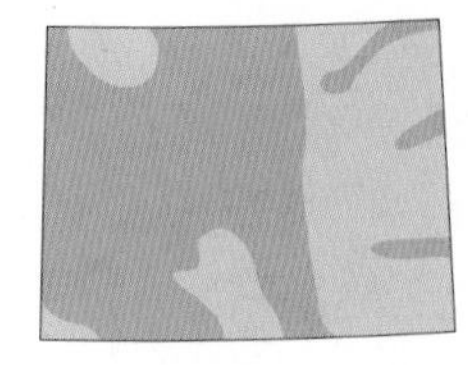

Behavior

The smallest woodpecker in North America, the Downy forages mainly on insects, larvae, and eggs. Readily visits backyard feeders for sunflower seeds and suet. Will also consume poison ivy berries. Small size enables it to forage on smaller, thinner limbs. Both male and female announce territorial claims with their drumming. Call is a high-pitched but soft *pik.*

Habitat

Common in suburbs, parks, and orchards, as well as forests and woodlands. Nests in cavities of dead trees.

Local Sites

Both Downy and Hairy Woodpeckers are found throughout wooded areas in Colorado; there is extensive overlap in the ranges of both. Both can be seen at bird feeders.

FIELD NOTES The larger Hairy Woodpecker, *Picoides villosus* (inset), is similarly marked but has a bill as long as its head and a sharper, louder, lower-pitched call. It also favors more mature woodlands, and tends to stay on tree trunks or larger limbs than the Downy. Note the red patch on the back of the head of the males of both species.

Year-round | Adult male "Red-shafted"

NORTHERN FLICKER

Colaptes auratus L 12.5" (32 cm)

FIELD MARKS

Brown, barred back; cream underparts with black spotting; black crescent bib; white rump

Brown crown and gray face

Red moustachial stripe on male, lacking on female

Behavior

The "Red-shafted" Northern Flicker is the common form seen in Colorado; its reddish orange feather shafts and wing linings most easily seen in flight. It feeds mostly on the ground, primarily on ants. Nests in cavities that it drills into wooden surfaces above ground, including utility poles and houses. Call is a long, loud series of *wick-er, wick-er* on breeding grounds, or a single, loud *klee-yer* year-round.

Habitat

Prefers open woodlands and suburbs with sizeable living and dead trees. An insectivore, it is partially migratory, moving southward in winter in pursuit of food.

Local Sites

A common migrant and winter resident in a variety of habitats, and a common summer resident of montane canyons and pine forests. Look for the "Red-shafted" Northern Flicker in wooded areas throughout the state.

FIELD NOTES The "Yellow-shafted" Northern Flicker (inset: male) is found in extreme eastern Colorado. It intergrades broadly with "Red-shafted" (opposite), and birds with intermediate characteristics are regularly seen.

Year-round | Adult

WESTERN WOOD-PEWEE

Contopus sordidulus L 6.25" (16 cm)

FIELD MARKS

Dark grayish brown overall; pale underparts

Broad, flat bill; yellow-orange at base of lower mandible

Long wings

Two thin white wing bars

Behavior

Solitary and often hidden in trees. As it perches, the Western looks actively about, without flicking tail or wings. When prey is spotted, it darts out to catch a variety of flies, spiders, butterflies, wasps, ants, and dragonflies. Often returns to the same perch. Bristly "whiskers" help it locate prey. Calls include a harsh, slightly descending *peeer,* and clear whistles suggestive of the Eastern Wood-Pewee's *pee-yer.* Song is heard chiefly on breeding grounds and has three-note *tswee-tee-teet* phrases mixed with the *peeer* note.

Habitat

Found in open woodlands, riparian areas, canyons, and along streams. Also in montane pine forests.

Local Sites

Common migrant and summer resident in most of Colorado, from spruce-fir forests to riparian areas such as Chatfield S. R. A. and Rocky Mountain N. P.

FIELD NOTES The Cordilleran Flycatcher, *Empidonax occidentalis* (inset), is Colorado's most commonly observed *Empidonax,* a genus of flycatchers which is notoriously challenging to identify. The Cordilleran regularly nests on cabins and other structures in the mountains, usually near water.

Year-round | Adult

SAY'S PHOEBE

Sayornis saya L 7.5" (19 cm)

FIELD MARKS

Brownish gray above, darkest on head, wings, and tail

Tawny buff belly and undertail coverts

Adult has indistinct pale gray wing bars; juvenile's are cinnamon

Behavior

The Say's Phoebe rarely stays in one spot for long. It darts around in pursuit of flying insects and wags its tail continually. Briefly perches on low-to-the-ground structures, such as branches, wires, and buildings. Song is often heard at dawn, consisting of two low-pitched, downslurred whistles, given alternately. Flight call is a quick *pit-tse-ar.* Typical call is a thin, whistled *pee-ee.*

Habitat

Unlike many flycatchers, the Say's Phoebe inhabits dry, treeless areas such as sagebrush plains, dry foothills and canyons, and dry farmland. Uses mud pellets, plant material, and spider webs to attach nest to the vertical plane of a rock or building.

Local Sites

Look for it in dry, open areas of Colorado. It is often seen around bridges and abandoned buildings.

FIELD NOTES The Black Phoebe, *Sayornis nigricans,* also hunts for flying insects from low perches, but is very much tied to habitats near water. It is easily distinguished from the Say's by its black body and white belly. Though its range is expanding, currently it is found only in the southwest corner of Colorado, and in an area west of Pueblo.

Year-round | Adult

WESTERN KINGBIRD

Tyrannus verticalis L 8¾" (22 cm)

FIELD MARKS

Ashy gray head, neck, breast

Back tinged with olive

Black tail with white sides

Lemon yellow belly

Behavior

Feeds almost exclusively on flying insects, leaving its perch to snatch prey in midair, often returning to the perch to eat. Perches horizontally instead of upright. Courtship display involves aerial flight. Like most kingbirds, builds cup-shaped nest near the end of a horizontal tree branch, lining it with weeds, moss, and feathers. Common and gregarious, nesting pairs regularly share the same tree. Call is a single or repeated sharp *kip;* also a staccato trill.

Habitat

Common in dry, open country. Has also adapted to more developed areas. Often perches on fence posts and telephone lines.

Local Sites

Very common summer resident and migrant, mainly at lower elevations throughout the state. Most common in open areas of eastern Colorado.

FIELD NOTES Cassin's Kingbird, *Tyrannus vociferans* (inset), a common summer resident in southeast Colorado, has darker gray upperparts than the Western, and a contrasting white chin. Call is a loud *chi-bew.*

Year-round | Adult

EASTERN KINGBIRD

Tyrannus tyrannus 8.5" (22 cm)

FIELD MARKS

Black head

Slate gray back

Tail has broad white terminal band

White underparts

Pale gray wash across breast

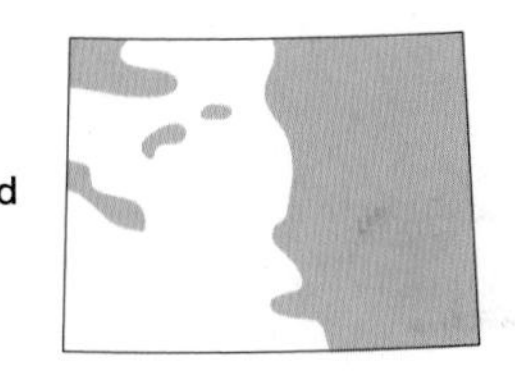

Behavior

Waits on perch until it sees an insect, catches prey in midair, returns to perch to eat. Males court with erratic hovering, swooping, and circling, revealing otherwise hidden orange-red crown patch. Builds cup-shaped nest of weeds, moss, and feathers near the end of a horizontal tree branch, sometimes on a post or stump. Raspy call when feeding or defending its territory sounds like *zeer;* also uses a harsh *dzeet* note alone or in a series.

Habitat

Common and conspicuous in woodland clearings or open fields with small forest stands, farms, and orchards.

Local Sites

Most easily seen in the eastern third of the state from May through September. Barr Lake State Park is a good place to see both Eastern and Western Kingbirds.

FIELD NOTES The Eastern Kingbird has an orange-red crown patch that is rarely seen unless the bird feels threatened or is displaying, raising the feathers on its head into a crest. Livingup to its Latin name, which means "tyrant of tyrants," the Eastern Kingbird will actively defend its nest, sometimes pecking at and even pulling feathers from the backs of hawks, crows, and vultures.

Year-round | Adult

LOGGERHEAD SHRIKE

Lanius ludovicianus L 9" (23 cm)

FIELD MARKS

Thick, black, finely hooked bill.

Wide black mask

Grayish body

Black wings and tail with white patches on wings and white outer tail feathers

Behavior

Shrikes are predatory songbirds that use their strong hooked bills to kill a variety of insects, small birds, mammals and reptiles. When hunting, shrikes perch conspicuously on fence lines, utility lines, or tops of trees and shrubs. They sometimes place prey on thorns or barbed-wire fences, which helped earn them the parochial name butcherbird.

Habitat

Prefers open country, particularly shortgrass prairie and open agricultural areas. Loggerhead Shrikes usually nest in isolated groves of trees or shrubs. Many abandoned farmsteads in eastern Colorado support nesting shrikes.

Local Sites

A year-round resident of southeastern Colorado, the Loggerhead Shrike is seen throughout the lowlands of most of Colorado. Pawnee National Grasslands is a good bet in summer, while Comanche National Grasslands is good all year.

FIELD NOTES The Northern Shrike is the most likely shrike to be seen from November through March in all but southeastern Colorado. The Northern Shrike has a narrower dark mask, longer, more hooked bill, and paler upperparts.

Year-round | Adult

PLUMBEOUS VIREO

Vireo plumbeus L 5.25" (13 cm)

FIELD MARKS

Prominent white spectacles and white throat; heavy bill

Two white wing bars

Gray upperparts; white below

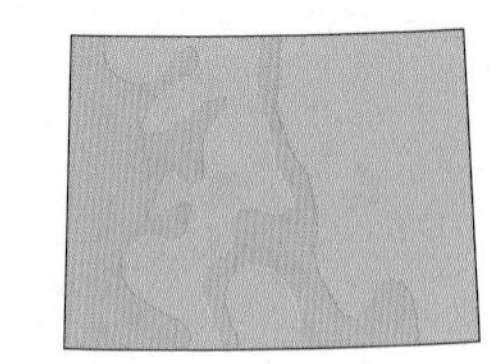

Behavior

Eats mainly insects, which it gleans from twigs and foliage, foraging in a slow, deliberate manner. Sometimes hovers and catches insects on the fly. Will also eat some fruit. Cup-shaped nest is usually built in a coniferous tree. Both parents incubate three to five eggs for up to two weeks. Hoarse song of burry two- and three-syllable phrases.

Habitat

Riparian woodlands, montane conifers, and mixed forests.

Local Sites

Common summer resident in lower montane canyons and pine forests in all the mountain ranges of the state. Fairly common migrant elsewhere. Common in the foothills in Rocky Mountain National Park, Castlewood Canyon, and throughout much of western Colorado.

FIELD NOTES The Gray Vireo, *Vireo vicinior* (inset), is a summer resident in pinyon-juniper habitat across western Colorado. It is also gray above and white below, but look for its single, thin wing bar and narrow white eye ring. Colorado National Monument is one of the best places to see this subtly marked species.

Year-round | Adult

STELLER'S JAY

Cyanocitta stelleri L 11.5" (29 cm)

FIELD MARKS

Black head, crest, and bill; white feathers above eye and on forehead

Dark gray back, neck, and breast

Purplish blue upperparts; smoky-blue underparts

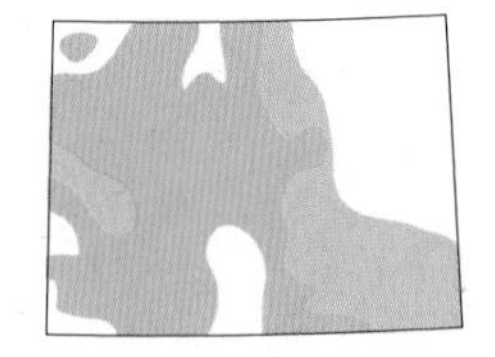

Behavior

Bold and aggressive. Regularly seen in flocks and family groups, feeding from the treetops to the ground. Often scavenges in campgrounds and picnic areas. Powerful bill efficiently handles a varied diet. Forages during warm months on insects, frogs, carrion, young birds, and eggs. Winter diet is mainly acorns and seeds. Hides food for later consumption. Highly social, Steller's Jays will stand sentry, ready to mob predators, while others in the flock forage. Calls include a series of *shack* or *shooka* notes.

Habitat

Common in pine-oak woods and coniferous forests, mainly at higher elevations.

Local Sites

Common throughout the Rocky Mountains in Colorado. Easily seen in Rocky Mountain National Park and other mountain locales.

FIELD NOTES The Blue Jay, *Cyanocitta cristata* (inset), the most abundant jay in North America, is found year-round throughout eastern Colorado, including the Denver metropolitan area.

Year-round | Adult

WESTERN SCRUB-JAY

Aphelocoma californica L 11" (28 cm)

FIELD MARKS

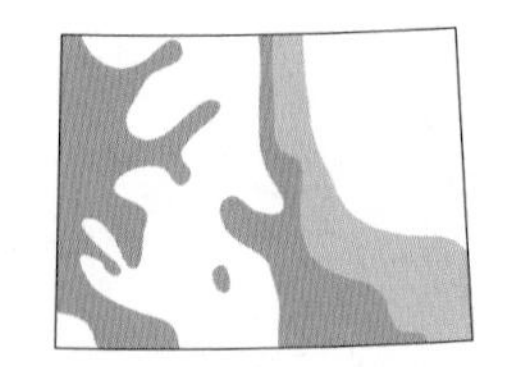

Dark blue upperparts; brownish gray back; grayish underparts

White eyebrow; dark eye patch

Whitish, streaked throat

Variable bluish band on chest

Juvenile has sooty gray hood

Behavior

Seen in pairs and small flocks, foraging on ground and in trees for insects, fruit, seeds, grain, eggs, young birds, and small reptiles and mammals. The Western Scrub-Jay's strong, stout bill allows for a wide-ranging diet. During courtship, male hops around the female with his tail spread and dragging on the ground. Both sexes build nest in tree or shrub using twigs, grass, and moss. This bold bird is not shy around humans and will often emit its metallic call, a raspy *shreep*, in a short series.

Habitat

Having increased its range significantly over the past several decades, this species can now be found in the scrublands for which it is named, as well as in deciduous woodlands and suburban areas.

Local Sites

Common and hard to miss in the foothills and valleys of the Front Range and western Colorado.

FIELD NOTES The Pinyon Jay, *Gymnorhinus cyanocephalus* (inset), travels in large, roaming flocks and is usually detected by its far-carrying calls.

Year-round | Adult

CLARK'S NUTCRACKER

Nucifraga columbiana L 12" (31 cm)

FIELD MARKS

Brownish gray body with whitish face; long black bill

Black wings; white patch on secondaries

Black and white tail with white undertail coverts

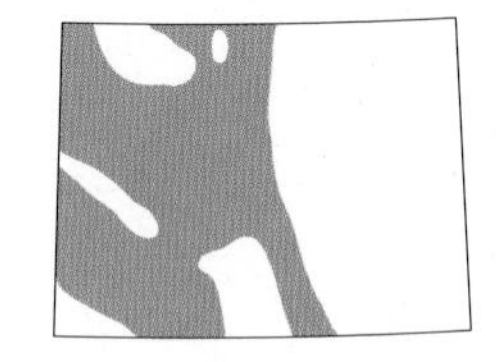

Behavior

The life cycle of Clark's Nutcracker is heavily dependent upon the availability of large seeds from several species of pines—in Colorado, primarily pinyon and limber pines. The year-round diet is based primarily on the seed from these cones, which the nutcracker transports to cache sites in a pouch below its tongue. Both sexes share brooding duty, permitting one parent to incubate the eggs while the other retrieves cached seeds. A variety of calls includes a long rising *shraaah.*

Habitat

Frequents coniferous forests, especially with pinyon and limber pines. Individuals and pairs are regularly seen at feeders and campsites, or flying over other habitats.

Local Sites

Clark's Nutcrackers are fairly common and widespread throughout much of Colorado's coniferous forest. Rocky Mountain National Park and Mount Evans are two locations close to Denver, but this species can also be found in most mountain towns and ski resorts.

FIELD NOTES During years of poor cone crops, nutcrackers may stage irruptions, when individuals may be seen along the Front Range and West Slope, outside of appropriate habitat. During such irruptions, individuals have strayed as far as the Midwest.

Year-round | Adult

BLACK-BILLED MAGPIE

Pica hudsonia L 19" (48 cm)

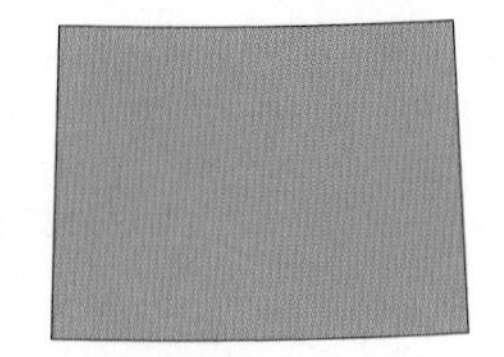

FIELD MARKS

Black upperparts, breast, and undertail coverts

White sides and belly

White wing patches show in flight

Bluish iridescence on wings; greenish iridescence on long tail

Behavior

Forages on the ground in family flocks of up to a dozen birds, feeding on insects, larvae, carrion, and sometimes on the eggs of other birds. In late summer and early fall, flocks may number 50 or more birds. Hoards food for later consumption and is also known to collect shiny nonfood items such as aluminum foil, glass shards, and even silverware. Typical calls include a whining *mag* and a series of loud, harsh *chuck* notes.

Habitat

Inhabits open woodlands, thickets, rangelands, and agricultural areas. Nests and roosts in riparian areas with dense overgrowth.

Local Sites

Magpies are widespread, common, gregarious, and resident in every Colorado county.

FIELD NOTES The Gray Jay, *Perisoreus canadensis* (inset), is resident in high-elevation spruce-fir forests. Frequent visitors to cabins, tent sites, and picnic grounds, where they brazonly beg for food.

Year-round | Adult

AMERICAN CROW

Corvus brachyrhynchos L 17.5" (45 cm)

FIELD MARKS

Black, iridescent plumage overall

Broad wings; fan-shaped tail

Long, heavy, black bill

Brown eyes

Black legs and feet

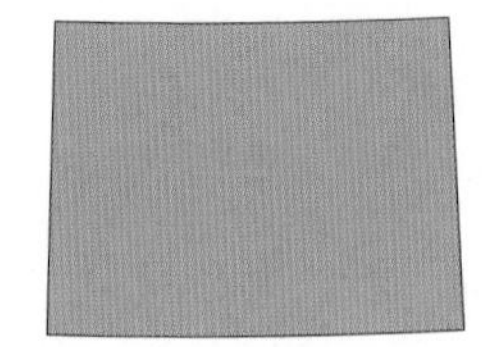

Behavior

Often forages, roosts, and travels in flocks. Individuals take turns at sentry duty while others feed on insects, garbage, grain, mice, eggs, and young birds. Known to noisily mob large raptors such as eagles, hawks, and Great Horned Owls, to drive them from its territory. Because its bill is ineffective on tough hides, crows wait for another predator—or an automobile—to open a carcass before dining. Studies have shown the crow's ability to count, solve puzzles, and retain information. Readily identified by its familiar call, *caw-caw.*

Habitat

One of North America's most widely distributed and familiar birds, lives in a variety of habitats, including urban areas. Nests in shrubs, trees, or on poles.

Local Sites

Common throughout its range, crows can be seen almost anywhere, except for high-elevation forests and open grasslands without trees.

FIELD NOTES American Crows form large flocks in late summer and early fall, but in Colorado such flocks are smaller than in parts of the East and Midwest where roosts of hundreds of thousands of birds are known.

Year-round | Adult

COMMON RAVEN

Corvus corax L 24" (61cm)

FIELD MARKS

Glossy black overall

Large, heavy bill with nasal bristles on top

Shaggy throat feathers

Wedge-shaped tail in flight

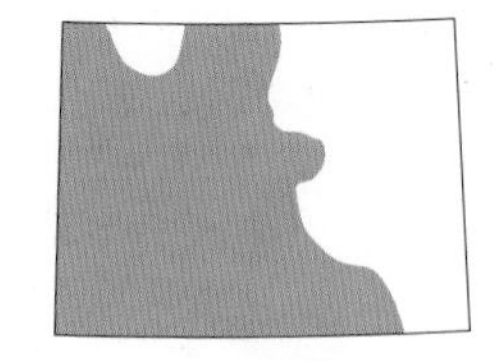

Behavior

The largest perching bird in North America, the Common Raven forages on a great variety of food, from worms and insects to rodents and eggs to carrion and refuse. Small groups are known to hunt together in order to overcome prey that is too large for just one bird to take. Monogamous for life, these birds engage in acrobatic courtship flights of synchronized dives, chases, and tumbles. Builds nest high in trees or on cliffs near water. Calls are extremely variable in pitch, from a deep croak to a high, ringing *tok.*

Habitat

The raven can be found in a variety of habitats, but is more abundant at higher elevations.

Local Sites

Common permanent resident at all elevations and in a variety of mountain habitats throughout Colorado.

FIELD NOTES Slightly smaller than the Common Raven, the extremely similar Chihuahuan Raven, *Corvus cryptoleucus,* is best identified by range and habitat. It is found in open country of southeastern Colorado. The base of the Chihuahuan's neck feathers is white, but this is very difficult to see in the field.

Year-round | Adult male

HORNED LARK

Eremophila alpestris L 6.75-7.75" (17-20 cm)

FIELD MARKS

Pale forehead bordered by black band, ending in hornlike tufts

Black cheek stripes

Pale yellow to white throat and underparts; brown upperparts

Sandy wash on sides and flanks

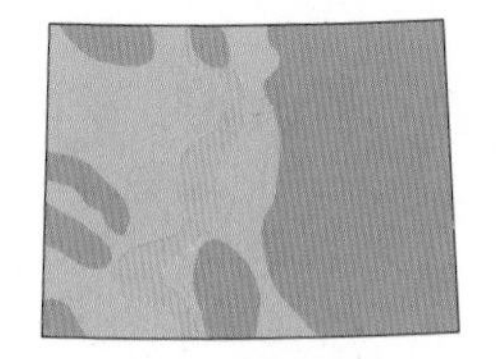

Behavior

Forages on ground, favoring open agricultural fields with sparse vegetation. Feeds mainly on seeds, grain, and some insects. Seldom alights on trees or bushes. On the ground, the Horned Lark walks rather than hops. Song is a weak twittering; calls include a high *tsee-ee* or *tsee-titi.* Females build the nest and incubate the clutch; both sexes feed the young. Outside breeding season, the birds form large flocks that often number in the hundreds.

Habitat

Found in dirt fields, gravel ridges, and grasslands. Also nests in alpine tundra.

Local Sites

Fairly common and widespread summer resident in open grassland habitats across the state. Locally abundant migrant and winter resident in agricultural areas. Look for large wintering flocks in fields in all of the lowland agricultural valleys in eastern Colorado.

FIELD NOTES The only native lark in North America, the Horned Lark is widespread, with about 20 subspecies (over 40 worldwide). Plumage tends to match the color of the soil it nests on.

Year-round | Adult male

VIOLET-GREEN SWALLOW

Tachycineta thalassina L 5.25" (13 cm)

FIELD MARKS

Dark above with green sheen on head and back; violet sheen on nape, wings, and tail

White below and on face, extending above eye

Pointed wings; notched tail

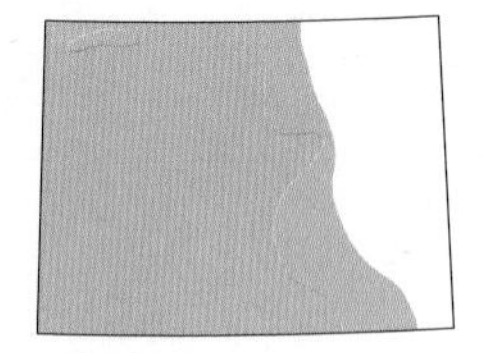

Behavior

Usually feeds in a flock on flying insects by darting close to the ground or low over water, but may also be seen hunting at greater heights. Perches in long rows high up in trees and on fences or wires. Call is a rapid, twittering *chi-chit;* song is a repeated *tsip-tsip-tsip,* most often given in flight around dawn.

Habitat

Found in various woodland settings and adjacent urban areas. Nests in dead trees, abandoned woodpecker holes, rock crevices, and man-made nest boxes. Diligently lines its nest with white feathers.

Local Sites

Common to abundant summer resident and migrant, particularly near water. Breeds at higher elevations in the Rocky Mountains. Rare in eastern Colorado away from the Rockies, even in migration.

FIELD NOTES The Tree Swallow, *Tachycineta bicolor* (inset: adult), which is also white underneath, is a more local breeder at high elevations and is also a very common migrant throughout. The Tree Swallow's white cheek patch does not extend above its eye, and its back is blue, not green.

Year-round | Adult

CLIFF SWALLOW

Petrochelidon pyrrhonata L 5.5" (14 cm)

FIELD MARKS

Blue-black crown, whitish forehead, chestnut throat and cheeks

Blue-black back with faint white streaks; whitish underparts

Squarish tail and buffy rump show in flight

Behavior

Primarily takes flying insects, but sometimes forages in shrubs or on ground for berries and fruit. Breeds in colonies of up to 1,000 birds in gourd-shaped nests (opposite) made of mud pellets, which the birds roll up in their bills and plaster to the sides of cliffs and buildings or underneath bridges and overpasses. Once Cliff Swallows depart for South America, wintering species readily take up the nests as roosting sites. Call is a low, husky *churr;* alarm call within the colony is a low *veer;* unmusical song consists of buzzes, squeaks, and rattles.

Habitat

Found in open riparian areas near cliffs, bridges, or buildings, but does not take well to urban settings.

Local Sites

Found throughout Colorado wherever there are nesting sites, from early April through late September.

FIELD NOTES The Northern Rough-winged Swallow, *Stelgidopteryx serripennis* (inset), is another summer breeder. It is drabber than the Cliff Swallow, dull brown above and white below, and it nests in burrows dug fairly deep into suitable soil, usually near water.

Year-round | Adult male

BARN SWALLOW

Hirundo rustica L 6.75" (17 cm)

FIELD MARKS

Long, deeply forked, dark tail

Iridescent deep blue upperparts; cinnamon to whitish underparts, paler on female

Rusty brown forehead and throat; dark blue-black breast band

Behavior

An exuberant flyer, the Barn Swallow is often seen in small flocks skimming low over the surface of a field or pond, taking insects in midair. Will follow tractors and lawn mowers to feed on flushed insects, many of which are harmful to crops. An indicator of coming storms, as barometric pressure changes cause the bird to fly lower to the ground. Call is a short, repeated *wit-wit.* Song is a husky warble interrupted by rattling creaks.

Habitat

Frequents open farms and fields, especially near water. Has adapted to humans to the extent that it now nests almost exclusively in structures such as barns, bridges, culverts, and garages. Widely distributed across the world, this bird also breeds in Europe and Asia and it winters in South America and southern Africa.

Local Sites

No barn, bridge, or culvert is complete without a couple of nesting pairs of Barn Swallows in summer.

FIELD NOTES Though the Barn Swallow nests in small colonies, competition between breeding pairs can be stiff. With its nest only inches from a neighboring nest, the male Barn Swallow vigorously defends his small territory. Unmated males have even been known to kill the nestlings of a mated pair in an attempt to break up the couple and mate with the female.

Year-round | Adult

BLACK-CAPPED CHICKADEE

Poecile atricapillus L 5.25" (13 cm)

FIELD MARKS

Black cap and bib

White cheeks

Grayish upperparts

Whitish underparts with rich buffy flanks, more pronounced in fall

Flight feathers edged in white

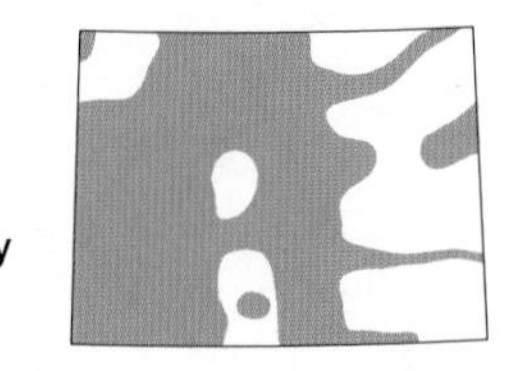

Behavior

A common backyard bird, the Black-capped Chickadee is often the first to find a new bird feeder. Also forages on branches and under the bark of various trees. Diet consists mostly of seeds, but eats some insects as well, and will hide food in different locations for later consumption. Calls include a low, slow, *chick-a-dee-dee-dee.* Song is a variable, clear, whistled *fee-bee* or *fee-beeyee,* the first note higher in pitch. Female at nest emits snakelike hissing if threatened.

Habitat

Common in open woodlands, clearings, and suburbs. Builds its nest of moss and animal fur in cavities in rotting wood or seeks out a man-made nest box.

Local Sites

The Black-capped Chickadee can be found in wooded suburban areas in much of Colorado.

FIELD NOTES The Black-capped Chickadee is the most widespread chickadee in North America. In Colorado, the highest densities are in aspen forests and along riparian corridors.

Year-round | Adult

MOUNTAIN CHICKADEE

Poecile gambelli L 5.25" (13 cm)

FIELD MARKS

White eyebrow differs from other chickadees—may be hard to see in summer

Black cap; black bib; white cheeks

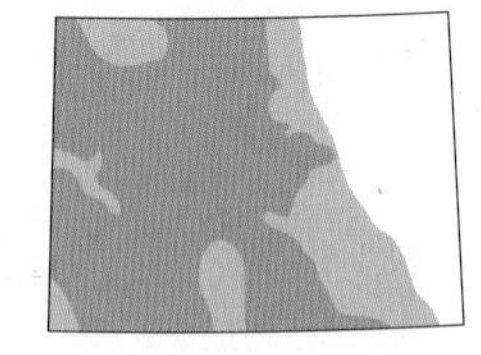

Behavior

The chickadee species most often seen in the higher mountains of the West. In fall and winter Mountain Chickadees forage in loose flocks with other forest songbirds for seeds, and insects hidden under bark, in pine needle clusters, and in the ground. Nest is a natural cavity or woodpecker hole; if no cavity is available, the bird digs a hole in decayed wood. Call is a hoarse chick-adee-adee-adee. Typical song is a three- or four-note descending whistle, fee-bay-bay or fee-bee, fee-bee.

Habitat

High elevation coniferous and mixed woodlands, particularly old-growth spruce-fir forests. In fall many individuals withdraw from the higher elevations and take up winter residence in lower-elevation mountain parks and foothills, and occasionally in western valleys and the eastern plains.

Local Sites

Most readily found in coniferous forests across the higher mountain ranges of Colorado.

FIELD NOTES Mountain Chickadees occasionally descend to the lowlands, where they can be found in the Denver metropolitan area and as far east as Kansas. Such irruptions usually take place from September to early April.

Year-round | Adult male

RED-BREASTED NUTHATCH

Sitta canadensis L 4.5" (11 cm)

FIELD MARKS

Blue-gray upperparts; rust-colored underparts

Black cap; white eyebrow; black postocular stripe; white cheeks

Female and juvenile have duller crown and paler underparts

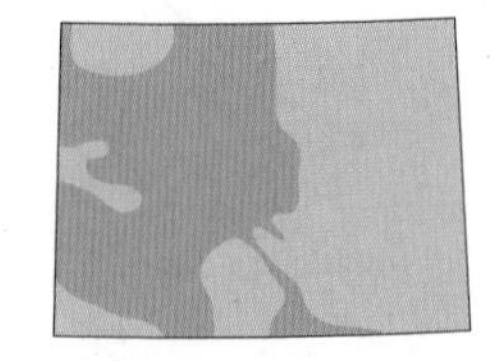

Behavior

Climbs up and down trunks, small branches, and outer twigs of trees, foraging for seeds, nuts, and insects. Known to join mixed-species foraging flocks during nonbreeding season and will often cache food. Male feeds female during courtship; display consists of male tilting side to side with head and tail upraised, back feathers pricked up, and wings dropped. Pairs smear pine pitch around nest entrance, presumably to ward off predators. High-pitched, nasal call is a repeated *ink.*

Habitat

Found in coniferous and mixed woods, but nests exclusively in conifers. Winter range varies each year, as this species is known to remain in its breeding range as long as food is available.

Local Sites

Any of Colorado's coniferous forests are likely to host breeding Red-breasteds in summer.

FIELD NOTES This small bird derives part of its name from its resourceful method of obtaining food. It will often wedge a nut or insect into a bark crevice, then hack it with its sharp, pointed bill until it breaks through the shell or exoskeleton and obtains the digestible parts within.

Year-round | Adult male

WHITE-BREASTED NUTHATCH

Sitta carolinensis L 5.75" (15 cm)

FIELD MARKS

Black crown on male, female's crown grayer

All-white face and breast; blue-gray upperparts; variable rust below

Thin, black bill, tip slightly upturned

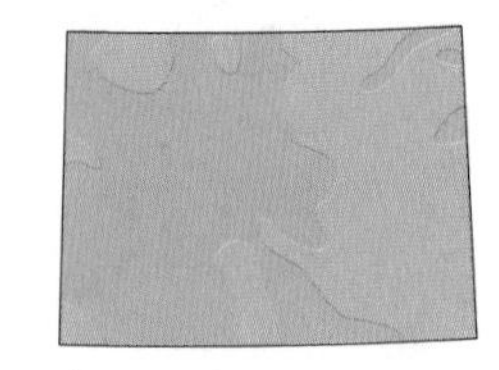

Behavior

An active, nimble feeder. Often spirals around a tree trunk, head down, foraging for insects in bark crevices. Readily visits backyard feeders, preferring sunflower seeds. The White-breasted Nuthatch in the Colorado region gives a very short, high-pitched, nasal call with a somewhat laughing quality—very different from eastern or Pacific coast birds. Notes are given in a rapid series, *nyeh-nyeh-nyeh-nyeh.*

Habitat

Prefers wooded areas full of oaks and conifers. Nests in woodpecker holes or natural cavities in decaying trees.

Local Sites

The White-breasted Nuthatch is a fairly common permanent resident in coniferous forests and montane riparian areas throughout the mountains of Colorado. Birds in the eastern subspecies group can be seen year-round at Bonny Lake State Park.

FIELD NOTES The White-breasted is the most widely distributed nuthatch in North America, and its call varies regionally. Eastern birds give a slow, low-pitched *yank;* Pacific birds are somewhat similar but give a higher pitched, longer, and harsher *eerh, eerh;* interior west birds are described above.

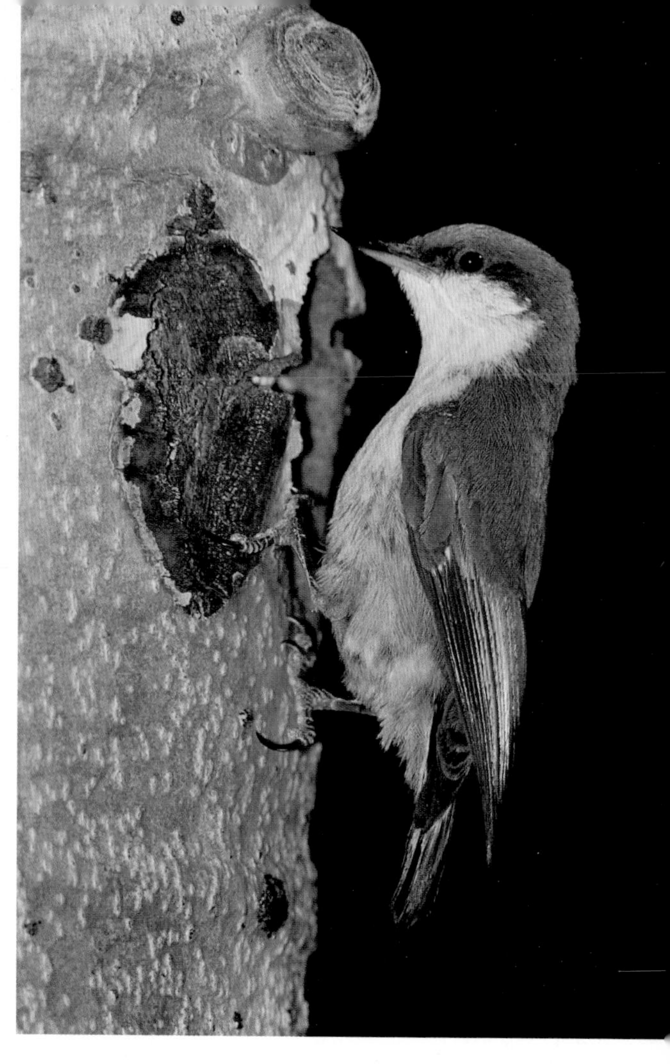

Year-round | Adult

PYGMY NUTHATCH

Sitta pygmaea L 4.25" (11 cm)

FIELD MARKS

Tiny; short-tailed

Blue-gray back; white to buff underparts

Gray-brown cap extends down to eyes

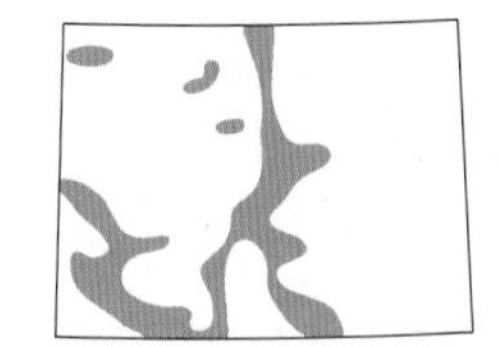

Behavior

Smallest nuthatch. Hops along tree trunks and branches and forages in clusters of pine needles at the tips of branches and in bark crevices; sometimes flies to the ground to feed. Hides and stores food. Roams in loose flocks. Both sexes excavate a small cavity, usually near the top of a dead pine or upright post. Female lays up to eight speckled reddish brown eggs, which hatch in about 16 days. Offspring from previous years help their parents raise the young. Typical calls include a high, rapid *peep-peep* and a piping *wee-bee.*

Habitat

Yellow and ponderosa pine forests.

Local Sites

Common in pine forests across Colorado, but greatly favors ponderosa pine forests. Pygmy Nuthatches can be easily found in the ponderosas at Rocky Mountain National Park.

FIELD NOTES The Pygmy Nuthatch is one of the few cooperative breeding songbirds. Pairs roost together and juveniles roost with parents. In winter, a dozen or more may join together in a single cavity.

Year-round | Adult

BROWN CREEPER

Certhia americana L 5.25" (13 cm)

FIELD MARKS

Mottled, streaky brown above

White eyebrow stripe

White underparts

Long, thin, decurved bill

Long, graduated tail

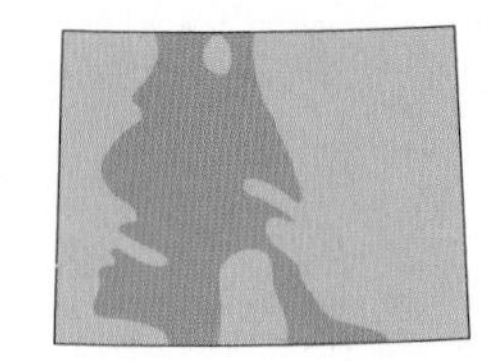

Behavior

Camouflaged by streaked brown plumage, creepers spiral upward from the base of a tree, then fly to a lower place on another tree in search of insects and larvae in bark crevices. A Brown Creeper will also eat fruit and berries when insects are scarce. Its long, decurved bill helps it to dig prey out of tree bark, its stiff tail feathers serving as a prop against the trunk. Forages by itself, in general, unless part of a mixed-species flock in winter. Call is a soft, sibilant, almost inaudible *see.* Song is a high-pitched *see-see-titi-see,* or a similar variation.

Habitat

Remains in forested areas, and builds nests behind loose bark of dead or dying trees. May wander into suburban and urban parks in winter.

Local Sites

The Brown Creeper can be difficult to spot, so listen for its high-pitched call in forested regions of the Rockies, or at lower elevations in fall and winter.

FIELD NOTES If the creeper suspects the presence of a predator, it will spread its wings and tail, press its body tight against the trunk of a tree, and remain completely motionless. In this pose, its camouflage plumage makes it almost invisible.

Year-round | Adult

CANYON WREN

Catherpes mexicanus L 5.75" (15 cm)

FIELD MARKS

Flat head and long bill; white throat and breast

Chestnut tinged body with fine speckles

Long, chestnut tail with thin black bars

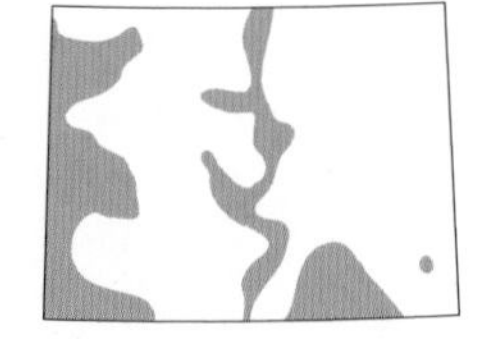

Behavior

Extracts insects from crevices in rock with long bill; occasionally flycatches. Can climb up, down, and across rocks. Pairs remain together throughout the year. Cup-shaped nest of twigs and grasses is built by both male and female. Female incubates five white eggs for 12 to 18 days, while the male feeds her. Loud, silvery song is a decelerating, descending series of liquid *tees* and *tews*. Typical call is a sharp *jeet*.

Habitat

Canyons, boulder piles, and cliffs, often near water.

Local Sites

Fairly common permanent resident near cliffs and rocky slopes in the mountains and canyons across the state. Look for the Canyon Wren, and listen for its echoing song, at Red Rocks Park and other areas with large cliffs.

FIELD NOTES The Rock Wren, *Salpinctes obsoletus* (inset), found in similar habitats, lacks the reddish-brown tones and contrasting white throat of the Canyon Wren and has a shorter bill.

Year-round | Adult

HOUSE WREN

Troglodytes aedon L 4.75" (12 cm)

FIELD MARKS

Grayish or brown upperparts

Fine black barring on wings and tail

Pale gray underparts

Pale faint eye ring and eyebrow

Thin, slightly decurved bill

Behavior

Noisy, conspicuous, and relatively tame, with a tail often cocked upward. Gleans insects and spiders from vegetation. While most species of wren forage low to the ground, the House Wren will seek food at a variety of levels, including high in the trees. Sings exuberantly in a cascade of bubbling, whistled notes. Call is a rough *chek-chek*, often running into a chatter.

Habitat

Inhabiting primarily open woodlands and thickets, this bird is also tolerant of human presence, and can be found in shrubbery around farms, parks, and urban gardens. Nests in any cavity of suitable size.

Local Sites

Forest openings, particularly at lower elevations, provide suitable nesting grounds for House Wrens in summer. One of Colorado's most abundant breeders.

FIELD NOTES Commonly hidden in the reedy marshes and cattail swamps of Colorado, the Marsh Wren, *Cistothorus palustris* (inset), reveals its location with its constant, abrasive call of *tsuk-tsuk*. Like the House Wren, it is brown overall, but has a prominent white eye stripe and its mantle is streaked black-and-white.

Year-round | Adult

AMERICAN DIPPER

Cinclus mexicanus L 7.5" (19 cm)

FIELD MARKS

Sooty gray overall

Short tail and wings

Straight dark bill

Pink legs and feet

Juvenile has paler, mottled underparts; pale bill

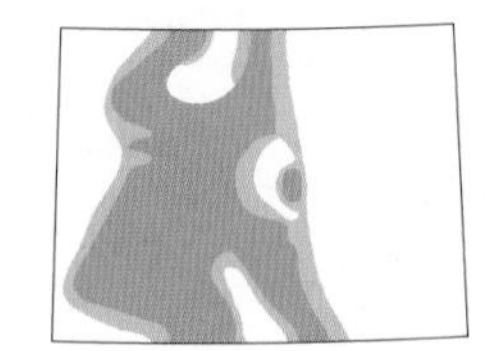

Behavior

Dippers are the only North American songbirds that swim. Using their wings to propel themselves underwater, they can walk on the river bottom to forage into crannies and under rocks for the larvae of various flies and mosquitoes. Also eat worms, water bugs, clams, snails, even small trout. In courtship, male will strut and sing with wings spread, after which both partners may jump up and bump breasts. Song is rattling and musical, loud enough to be heard over river's splash.

Habitat

Found along mountain streams fed by melting snow, glaciers, and rainfall. Descends to lower elevations in winter. Nests close to water level on cliffs, midstream boulders, bridges, or behind waterfalls.

Local Sites

Any fast-flowing, clear stream or river may host an American Dipper. Favorite locations include Morrison Park and along rivers in Rocky Mountain National Park.

FIELD NOTES The American Dipper has the uncanny ability to fly directly into and out of water. It may sometimes wade in from a river bank and dive from there, but it can also dive straight into water from a low flight.

Year-round | Adult male

RUBY-CROWNED KINGLET

Regulus calendula L 4.25" (11 cm)

FIELD MARKS

Olive green above; dusky below

Yellow-edged plumage on wings

Two white wing bars

Short black bill; white eye ring

Male's red crown patch seldom visible except when agitated

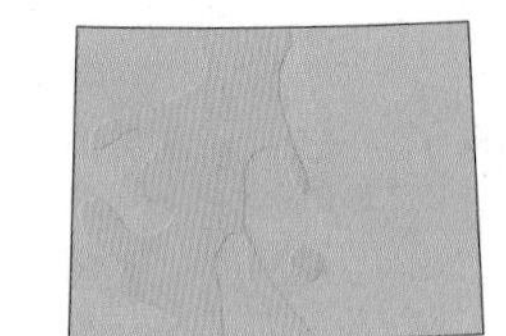

Behavior

Often seen foraging in mixed-species flocks, the Ruby-crowned Kinglet flicks its wings frequently as it searches for insects and their eggs or larvae on tree trunks, branches, and foliage. May also give chase to flying insects or drink sap from tree wells drilled by sapsuckers. Calls include a scolding *ji-dit;* song consists of several high, thin *tsee* notes, followed by descending *tew* notes, ending with a trilled three-note phrase.

Habitat

Common in coniferous and mixed woodlands, brushy thickets, and backyard gardens. Highly migratory.

Local Sites

Colorado's most common nesting bird in coniferous forests, from late April. A few winter in Colorado, along the Arkansas River near Pueblo, also near Grand Junction and Delta.

FIELD NOTES Often in the company of the Ruby-crowned, the Golden-crowned Kinglet, *Regulus satrapa* (inset), can be found in winter as it forages high up in trees. It is set apart by its yellow crown patch and its striped head. The male (inset, bottom) shows a brilliant orange tuft within his yellow crown patch.

Breeding | Adult male

BLUE-GRAY GNATCATCHER

Polioptila caerulea L 4.25" (11 cm)

FIELD MARKS

Male is blue-gray above, female grayer; both are white below

Long, black tail with white outer feathers

Black forehead and eyebrow on male in breeding plumage

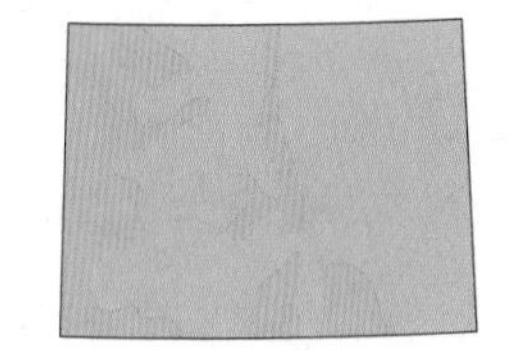

Behavior

Often seen near branch tips, the gnatcatcher scours deciduous tree limbs and leaves for small insects, spiders, eggs, and larvae. Sometimes captures prey in flight and may hover briefly. Male and female together make cuplike nest of plant fibers, spider webs, moss, and lichen on a branch or fork of a tree. Distinguished by its high-pitched buzz while feeding or breeding. Also emits a querulous *pwee*, intoned like a question. Known to imitate other birds' songs.

Habitat

Favors moist woodlands and thickets.

Local Sites

Breeding pairs of Blue-gray Gnatcatchers can be found along the Front Range and throughout most of the West Slope. Look for it in foothills outside Denver, such as Gregory Canyon and Red Rocks Park.

FIELD NOTES Similar to a gnatcatcher in shape but duller in plumage, the Bushtit, *Psaltriparus minimus* (inset: male), is fairly common in much of western and southern Coorado, It travels in small roving flocks, usually in pinyon-juniper woodlands, or along riparian stretches in winter—and is increasingly seen in residential areas.

Year-round | Adult male

WESTERN BLUEBIRD

Sialia mexicana L 7" (18 cm)

FIELD MARKS

Chestnut shoulders, upper back

Deep purple-blue upperparts and throat in male; duller, brownish gray in female

Chestnut breast, sides, and flanks in male; chestnut gray in female

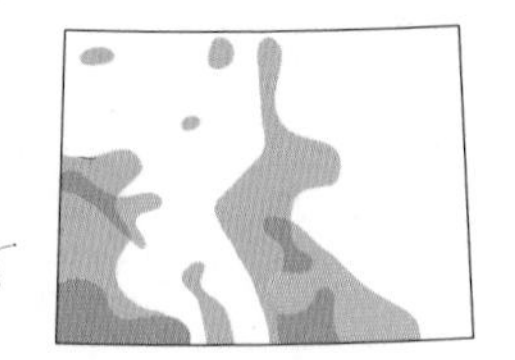

Behavior

Using large, acute eyes, a bluebird will hunt from a perch high above the ground, swoop down to seize crickets, grasshoppers, and spiders, which it may have spotted from as far away as 130 feet. The call note of the Western Bluebird is a mellow *few,* extended in brief song to *few few fawee.*

Habitat

Found in woodlands, farmlands, and orchards; and deserts in the winter. Nests in holes in trees or posts as well as in nest boxes. Frequents mesquite-mistletoe groves in winter.

Local Sites

Most commonly seen in southwestern Colorado, but also regularly seen north along the Front Range to Rocky Mountain National Park.

FIELD NOTES In general, the Western Bluebird prefers more wooded habitat than the Mountain Bluebird (p. 176), but both species may be seen together, particularly in winter.

Year-round | Adult male

MOUNTAIN BLUEBIRD

Sialia currocoides L 7.25" (18 cm)

FIELD MARKS

Male is a bright sky blue overall, paler below

Female is dusky blue-gray overall, sometimes washed with rusty orange on the breast

Longer-winged than Western Bluebird

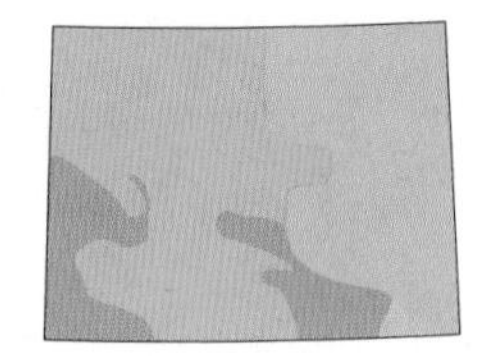

Behavior

Generally seen alone, in a pair, or in a small family group. Hunts from a perch for flying insects, caterpillars, and grasshoppers. Also known to hover over grasslands while searching for prey. May forage on the ground for fruit and berries, particularly in winter. Nests in tree cavities or nest boxes. Its call is a thin *few;* its song is a series of low, warbled *tru-lee*s.

Habitat

Found in open woodlands, agricultural areas, and grasslands generally at fairly high elevations. Highly migratory, tends to abandon higher elevations as winter progresses.

Local Sites

The Mountain Bluebird is easily found in open areas in the mountains. Hard to miss at Rocky Mountain National Park in summer.

FIELD NOTES Female Mountain Bluebirds are less striking than adult males, with a subdued grayish body and wash of blue on the wings and tail. Some females have a rufous or buff wash to the breast, which sometimes leads to confusion with Western Bluebirds (p. 174), or Eastern Bluebirds (*Sialia sialis*).

Year-round | Adult

TOWNSEND'S SOLITAIRE

Myadestes townsendi L 8.5" (22 cm)

FIELD MARKS

Gray overall; long tail

White eye ring

White outer tail feathers

Intricate buff-colored patterning to wings with buffy wing stripe visible in flight.

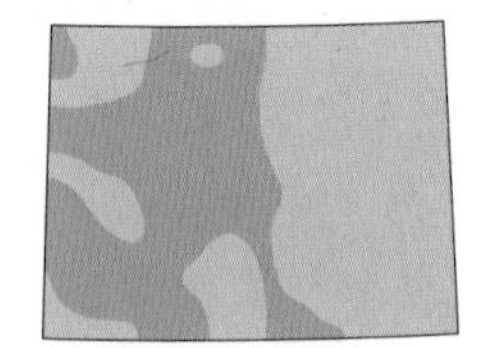

Behavior

Usually found alone, a solitaire will often perch conspicuously, usually at the top or near the top of a tree. Solitaires build cup nests made of pine needles and lined with grass, placed on the ground or on a cliff sheltered by an overhang. The frequently heard call note is a plaintive clear whistled *heeh,* that is sometimes confused with a single toot of a pygmy-owl. The song is a series of finchlike warbles with little pattern.

Habitat

Breeds in all types of coniferous forest in Colorado, preferring relatively open habitats to dense stands. In winter, many descend to the lowlands, where they can be found in almost any habitat with shrubs and trees.

Local Sites

A widespread species in Colorado, the Townsend's Solitaire is usually easy to find in coniferous forest where there are rocky cliffs or other suitable habitats on which to build nests. In winter, many descend to the lowlands and may be seen anywhere in Colorado, usually feeding in junipers or ornamental fruiting trees and shrubs.

FIELD NOTES Juvenile Townsend's Solitaires are very dark with an intricate pattern of scales and spots. They still have the characteristic solitaire shape, with a proportionately very long tail and at least a trace of the adult wing pattern.

Year-round | Adult

HERMIT THRUSH

Catharus guttatus L 6.75" (17cm)

FIELD MARKS

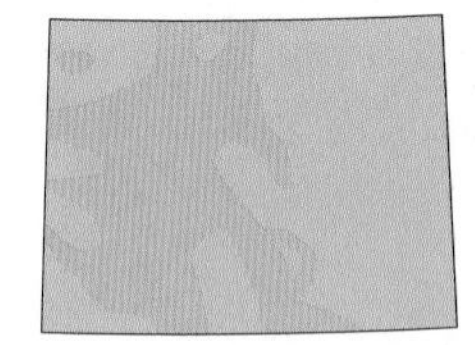

Gray-brown upperparts; white to buffy underparts with dense spotting mostly on breast

Reddish tail contrasts with upperparts

Whitish eye ring; dark malar stripe

Behavior

The Hermit Thrush is a shy, terrestrial bird that forages on the forest floor for insects or ascends into bushes in search of berries. When interrupted it flies up into a low bush, flicking its wings nervously and slowly raising and lowering its tail. Common call is a blackbird-like *chuck*, often doubled; song is a serene series of clear, flutelike notes, the phrases repeated at different pitches, lending it a lyrical quality.

Habitat

For breeding, coniferous forests typically in areas of relatively little undergrowth. In winter and migration uses a wide variety of habitats.

Local Sites

Fairly common summer resident in mountains. While rare in winter, the Hermit Thrush is the only brown woodland thrush (*Catharus*) likely to be seen in winter.

FIELD NOTES Similar to the Hermit thrush, the Swainson's Thrush, *Catharus ustulatus* (inset), has a more conspicuously bold, buffy eye ring and less reddish tail. Swainson's is the most common *Catharus* thrush in migration.

Year-round | Adults

AMERICAN ROBIN

Turdus migratorius L 10" (25 cm)

FIELD MARKS

Brick red underparts, paler in female, spotted in juvenile

Brownish gray above with darker head and tail

White throat and lower belly

Broken white eye ring; yellow bill

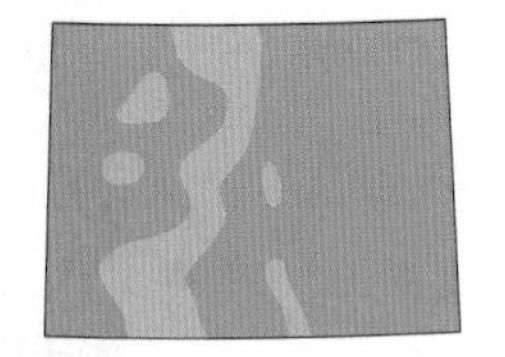

Behavior

Best known and largest of the thrushes, often seen on suburban lawns, hopping about and cocking its head in search of earthworms. The American Robin gleans butterflies, damselflies, and other flying insects from foliage and sometimes takes prey in flight. Robins also eat fruit, especially in fall and winter. This broad plant and animal diet makes them one of the most successful and wide-ranging thrushes. Calls include a rapid *tut-tut-tut;* song is a variable *cheerily cheer-up cheerio.*

Habitat

Common and widespread, the American Robin forages on lawns and in woodlands. Nests in shrubs, trees, and even on sheltered windowsills. Winters in moist woodlands, suburbs, and parks.

Local Sites

Look for robins year-round almost anywhere in Colorado, including the nearest backyard. This adaptable bird thrives in residential, agricultural, and natural habitats.

FIELD NOTES The juvenile robin, which can be seen between May and September, has a paler breast, like the female of the species, but its underparts are heavily spotted with dark brown. Look as well for the buff fringes on its back and wing feathers and its short, pale buff eyebrow.

Year-round | Adult

SAGE THRASHER

Oreoscoptes montanus L 8.5" (22 cm)

FIELD MARKS

Heavily streaked underparts

Long tail with white corners

Pale iris; relatively short bill

By late summer, streaking and white wing bars often fainter

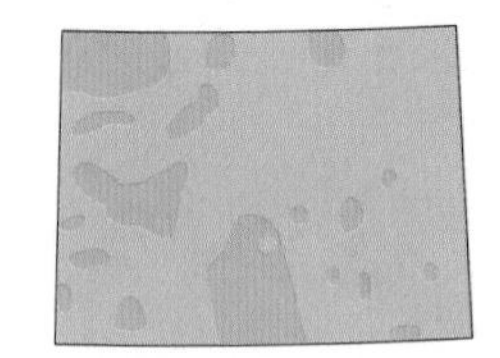

Behavior

The most widespread of Colorado's thrashers, the Sage Thrasher usually forages near the ground for a variety of invertebrates. It will also feed on small fruits when available. When disturbed, Sage Thrashers will often run rather than fly. The song is rich and melodious and can last for several minutes, the singing bird usually perched conspicuously near the tops of shrubs.

Habitat

Aptly named, the Sage Thrasher favors breeding in areas with extensive sage and other shrubs, especially in mountain valleys and plains. During migration individuals can be found in almost any habitat, regularly dispersing in late summer up to alpine tundra.

Local Sites

Common in summer in western Colorado's mountain parks and valleys, where sage and other short shrubs are abundant.

FIELD NOTES In addition to the catlike *mew* that gives the species its name, the Gray Catbird, *Dumetella carolinensis* (inset), can reproduce the calls of other birds, amphibians, and machinery, which it incorporates into its song. It is a fairly common breeder along riparian stretches in most of Colorado, a common migrant in eastern Colorado.

Nonbreeding | Adult

EUROPEAN STARLING

Sturnus vulgaris L 8.7" (22 cm)

FIELD MARKS

Iridescent black in spring, summer

Buffy tips on back, tail feathers

Fresh fall feathers tipped in white, giving speckled appearance

In summer, base of yellow bill is pale blue on male, pink on female

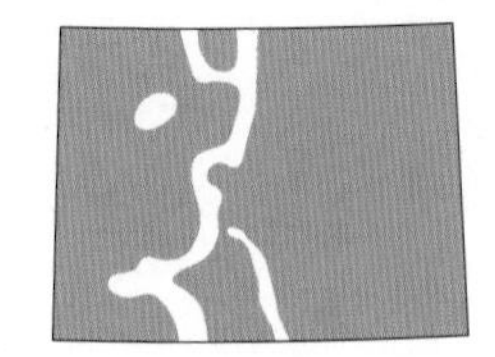

Behavior

A social and aggressive bird, the European Starling feeds on a tremendous variety of food, ranging from invertebrates—such as snails, worms, and spiders—to fruits, berries, grains, seeds, and garbage. It probes the ground for food, opening its bill to create small holes and expose prey. Usually seen in flocks, except during nesting season. Imitates calls of other species, especially grackles, and emits high-pitched notes, including squeaks, hisses, chirps, and twittering.

Habitat

The adaptable starling thrives in a variety of habitats near humans, from urban centers to agricultural regions. Nests in cavities, ranging from crevices in urban settings to woodpecker holes and nest boxes.

Local Sites

Widespread year-round throughout Colorado, the starling is likely to be found in most local parks.

FIELD NOTES A Eurasian species introduced into New York's Central Park in 1890, the European Starling has since spread throughout the U.S. and Canada. Abundant, bold, and aggressive, starlings often compete for and take over nest sites of other birds, including bluebirds, Wood Ducks, a variety of woodpeckers, Tree Swallows, and Purple Martins.

Year-round | Adult

CEDAR WAXWING

Bombycilla cedrorum L 7.3" (19 cm)

FIELD MARKS

Distinctive, sleek crest

Black mask bordered in white

Silky plumage with brownish chest and upperparts

Yellow terminal tail band

May have red, waxy tips on wings

Behavior

Eats the most fruit of any bird in North America. Up to 84 percent of its diet includes cedar, peppertree, and hawthorn berries and crabapple fruit. Also consumes sap, flower petals, and insects. Cedar Waxwings are gregarious in nature and band together for foraging and protection. Flocks containing from a few to thousands of birds may feed side by side in winter. Flocks may rapidly disperse, startling potential predators. Call is a soft, high-pitched trilled whistle.

Habitat

Found in open habitats where berries are available. The abundance and location of berries influence the Cedar Waxwing's migration patterns; it will move long distances only when its food sources run out.

Local Sites

Locally common yet irregular migrant, breeder, and winter resident across Colorado.

FIELD NOTES The only other North American member of the Bombycillidae family is the Bohemian Waxwing, *Bombycilla garrulus* (inset). Larger and grayer than the Cedar, its wings are intricately marked in white, black, red, and yellow. Look for flocks between December and February wherever fruit-bearing trees occur in Colorado, sometimes with flocks of Cedar Waxwings.

Year-round | Adult

VIRGINIA'S WARBLER

Vermivora virginiae L 4.8" (12 cm)

FIELD MARKS

Yellow rump, uppertail coverts and undertail coverts

Prominent white eye ring

Grayish body and wings

Yellow patch on breast, weak or absent in juvenile

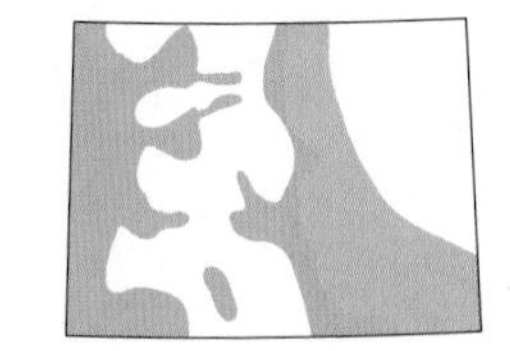

Behavior

Gleans insects and larvae from the branches and vegetation of pines, oaks, and other trees. In migration sometimes forages in small flocks with other warblers, chickadees, and nuthatches. The male's song is a fast trill, with the last notes on a lower pitch. The short call note is a sharp high *chip.*

Habitat

A common breeder in Gambel's oak, mountain mahogany, and other shrubby trees. During migration, most Viriginia's are still found in shrubby vegetation, usually within a foot of the ground.

Local Sites

Easily found at locations such as Colorado National Monument and Red Rocks Park. Generally scarce on the far eastern plains in migration, but regularly seen elsewhere in the state as a migrant.

FIELD NOTES The Orange-crowned Warbler, *Vermivora celata* (inset), is similar but more yellowish overall with a thin dark eye line and blurry olive streaks on the sides of its breast. They are common breeders in the west and common migrants throughout Colorado.

Year-round | Adult male

YELLOW WARBLER

Dendroica petechia L 5" (13 cm)

FIELD MARKS

Bright yellow overall

Plump and short-tailed

Dark eye prominent in yellow face

Male shows distinct reddish streaks below; streaks faint or absent in female

Behavior

Forages in trees, shrubs, and bushes, gleaning insects, larvae, and fruit from their branches and leaves. Will sometimes spot flying insects from a perch and chase them down. Mostly seen singly or in a pair. Nests in the forks of trees or bushes at eye level or a little higher. Male and female both feed nestlings, sometimes mistakenly giving them noxious, leaf-eating caterpillars. Song is a rapid, variable *sweet-sweet-I'm-so-sweet.*

Habitat

Favors wet habitats, especially those with willows and alders, but also lives in open woodlands, gardens, and orchards.

Local Sites

The most widespread breeding warbler in Colorado. May be seen at almost any location with riparian habitat from May through early September.

FIELD NOTES A common victim of cowbird nest parasitism (p. 243), the Yellow Warbler has devised an interesting retaliation tactic. Once foreign eggs are detected, the female will build a new roof of grasses, moss, lichen, and fur over all the eggs, then simply lay a new clutch. A single nest has been found to have up to six stories embedded with cold cowbird and warbler eggs.

Breeding | Adult male "Audubon's"

YELLOW-RUMPED WARBLER

Dendroica coronata L 5.3" (13 cm)

FIELD MARKS

Bluish gray head; yellow crown patch and throat; white eye crescents

Yellow rump and flank patch; breeding male has black breast

Females and winter males browner and streaked below

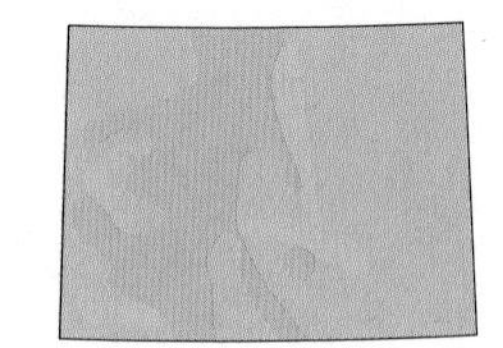

Behavior

The western subspecies group of the Yellow-rumped that breeds in Colorado is known as the "Audubon's Warbler"(see Field Notes). It darts from tree to tree among the branches, foraging for insects and berries; sallies from treetops to catch insects on the fly. Female lays four or five spotted white eggs in a bulky nest of twigs, roots, and grass lined with hair and feathers and built in a conifer. Song is a variable slow warble that rises or falls toward the end. Call is a loud, oft-repeated *chip* or *chep.*

Habitat

Coniferous or mixed woodlands.

Local Sites

Common summer resident at higher elevations in all of the major mountain ranges across the state. Locally abundant migrant in riparian areas. Smaller numbers winter in Grand Junction and Pueblo areas.

FIELD NOTES The eastern form of the Yellow-rumped Warbler, "Myrtle Warbler" (inset), is a common migrant in eastern Colorado. It has a white throat and pale supercilium rather than the yellowish throat and eye crescents of the "Audubon's."

Year-round | Adult male

MACGILLIVRAY'S WARBLER

Oporonis tolmiei L 5.3" (13 cm)

FIELD MARKS

Grayish hood

Bold white eye arcs

Black lores in male

Yellow underparts

Olive upperparts

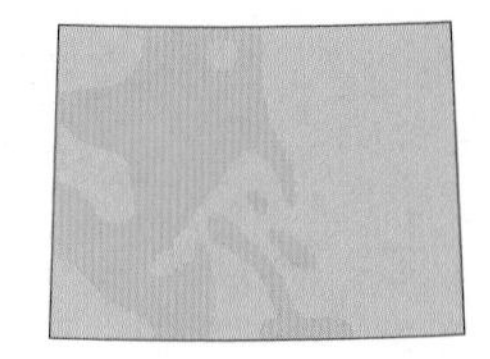

Behavior

A relatively secretive warbler that forages low to the ground, MacGillivray's are most often found by listening for their harsh dry *chik* call note, similar to that of the Common Yellowthroat (p. 198). Males are often most easy to detect singing from exposed branches and perches—a rich series of *churry* notes with the final phrases often buzzier and quieter.

Habitat

Prefers shrubby areas, nesting along mountain streams, but equally at home in regenerating clearcuts. Most common in Colorado along riparian habitats with moist understory. In migration, may be found throughout the state in dense tangles and shrubbery.

Local Sites

More birders have probably seen their first MacGillivray's Warbler in the wet thickets near Endovalley picnic area at Rocky Mountain National Park than in any other single location in Colorado.

FIELD NOTES While widespread as a breeder and migrant, the MacGillivray's Warbler can be difficult to find because of its secretive nature. Listen for its loud ringing song emanating from mountain streams, or for its dry, sharp *chik* call.

Year-round | Adult male

COMMON YELLOWTHROAT

Geothlypis trichas L 5" (13 cm)

FIELD MARKS

Adult male shows broad, black mask bordered above by light gray

Female lacks black mask, has whitish patch around eyes

Grayish olive upperparts; bright yellow throat and breast; pale yellow undertail coverts

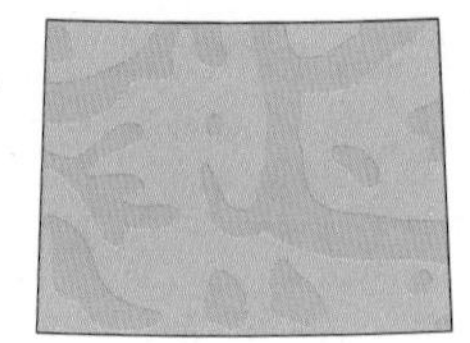

Behavior

A widespread warbler, the Common Yellowthroat generally remains close to the ground, skulking and hiding in undergrowth. May also be seen climbing vertically on stems and singing from exposed perches. While foraging, cocks tail and hops on ground to glean insects, caterpillars, and spiders from foliage, twigs, and reeds. Sometimes feeds while hovering, or chases flying insects. One version of variable song is a loud, rolling *wichity-wichity-wich.* Calls include a raspy *chuck.*

Habitat

Stays low in marshes, shrubby fields, woodland edges, and thickets near water. Nests atop piles of weeds and grass, or in small shrubs.

Local Sites

Found in summer and migration in wet habitats. Good locations include the Wheatridge Greenbelt, Barr Lake State Park, and Chatfield State Recreation Area.

FIELD NOTES The colors of the Common Yellowthroat vary considerably among subspecies separated by geography. Differences include the quality of yellow on the underparts, the extent of olive shading on the upperparts, and the width and color of the border between the male's mask and crown, which can range from stark white to gray.

Year-round | Adult male

WILSON'S WARBLER

Wilsonia pusilla L 4.8" (12 cm)

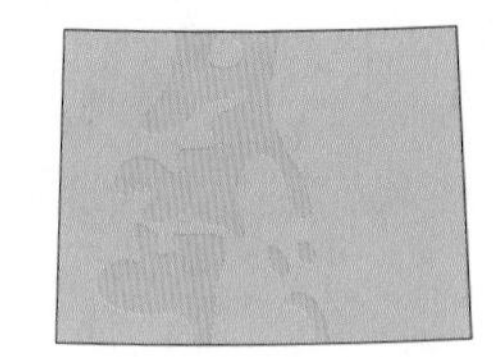

FIELD MARKS

Bright yellow underparts

Olive upperparts

Bright yellow supercilium

Yellow lores

Males have black crown

Behavior

A Wilson's Warbler can be found foraging at all heights, but is most frequently seen at or near eye level. Its tail is often raised and flipped from side-to-side as it forages for insects from foliage and twigs. It sometimes hovers to pick prey from leaves and will catch insects in flight. Its short, chattering song drops in pitch near the end; its frequently heard call note is a distinctive sharp, somewhat nasal *jimp.*

Habitat

The Wilson's Warbler favors moist habitats for breeding, including riparian stretches, beaver ponds, and overgrown montane clearcuts. Generally breeds at higher elevations than MacGillivray's Warbler. During migration can be found nearly anywhere, including backyard feeders.

Local Sites

Among the most common migrants, particularly in fall when they may be seen throughout the state. In summer, look in willow thickets along high mountain streams at such places as Rocky Mountain National Park.

FIELD NOTES One study in Colorado suggests that Wilson's Warblers in areas that are heavily used by recreational visitors may have a lower reproductive success due to higher rates of nest desertion than in more remote locales.

Year-round | Adult

YELLOW-BREASTED CHAT

Icteria virens L 7.5" (19 cm)

FIELD MARKS

Large size

Olive upperparts

Bright yellow throat, breast

White belly, undertail coverts

Dark lores contrast with white spectacles

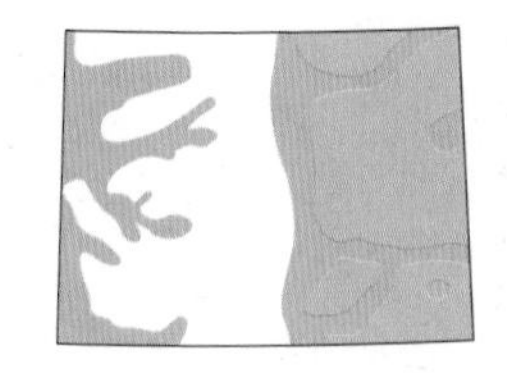

Behavior

Somewhat shy and difficult to see. Chats are most easily seen by listening for their distinctive song, which consists of a series of simple rich and loud low whistles, rattles and catcalls with long pauses between notes and phrases. Unusual for a bird of dense thickets, some songs are given in the air while the bird hovers. From late in the breeding season until they arrive again on the breeding grounds, the best clue to a chat's presence is the harsh call *cheewh.*

Habitat

Stays in dense second growth, riparian thickets and brush. In Colorado most are usually found near water. Their bulky cup nests are placed in dense shrubbery.

Local Sites

The Yellow-breasted Chat is a fairly widespread breeder and migrant in lowlands of eastern and western Colorado; more local at lower elevations in the mountains. Chatfield State Recreation Area is an excellent place to look for this species.

FIELD NOTES From its eclectic song, which is often given at night, to its large size, the Yellow-breasted Chat is perhaps the most distinctive wood-warbler.

Breeding | Adult male

WESTERN TANAGER

Piranga ludoviciana L 7.3" (19 cm)

FIELD MARKS

Bright red hood on breeding male

Yellow underparts, nape, and rump; yellow-green face on female

Black wings and tail

Upper wing bar is yellow, lower wing bar is white

Behavior

Forages both in trees and on ground for insects, especially wasps and bees, and for fruit. May join mixed-species foraging flocks after breeding. Known to visit birdbaths, but rarely feeders. Both sexes are known to sing a hoarse three-to-five-phrase series of repeated *chu-wee* notes, somewhat resembling American Robin or Black-headed Grosbeak. Call is a quick, slurred rattle: *pit-ick, pit-er-ick,* or *tu-weep.*

Habitat

Found in coniferous and pine-oak forests. Cup-shaped nest is located far out on branches.

Local Sites

The Western Tanager can be found in spring and summer in forests throughout much of Colorado. May also appear in any number of habitats during migration, including suburban areas.

FIELD NOTES Female (inset) and winter-plumaged Western Tanagers lack the breeding male's red head. Note the species' two wing bars and relatively large bill.

Year-round | Adult

GREEN-TAILED TOWHEE

Pipilo chlorurus L 7.3" (19 cm)

FIELD MARKS

Rufous crown

Bright yellow-green edging to wings and tail

Dark gray breast and gray belly

Bright white throat

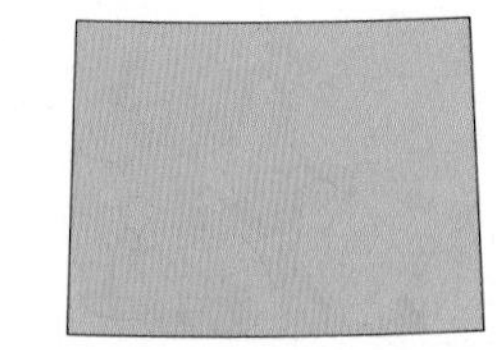

Behavior

The Green-tailed Towhee is one of Colorado's most widespread breeders, but can be somewhat shy and difficult to see when it is not singing. Towhees hop on the ground while foraging for seeds and small insects. Most foraging takes place amid dense foliage with surrounding and overhanging thickets and shrubs. The varied song usually consists of several introductory notes followed by a series of trills. The soft call is a distinctive nasal *meewee.*

Habitat

Found in a variety of dense brushy habitats, from stands of mountain mahogany and Gambel's oak to the understory of juniper woodlands.

Local Sites

Listen for its rich song and call notes at Red Rocks Park and throughout the shrubby terrain of the West Slope. It occasionally visits backyard feeders and bird baths.

FIELD NOTES Juvenile Green-tailed Towhees are shaped like adults but differ completely in plumage. They are heavily streaked above and below with a prominent white throat. They look similar to Spotted Towhees (p. 208), but their wings are browner, usually with greenish edging.

Year-round | Adult male

SPOTTED TOWHEE

Pipilo maculatus L 7.5" (19 cm)

FIELD MARKS

Black upperparts and hood on male; gray-brown on female

Rufous flanks and white underparts; white spots on back and scapulars; two white wing bars

Long tail with white corners

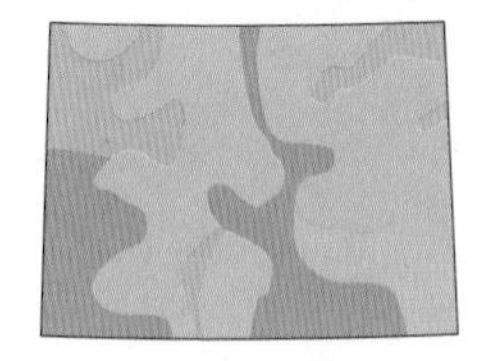

Behavior

This species employs the distinctive double-scratch technique—kicking its feet backward in the leaf litter, head held low and tail pointed up, attempting to expose seeds, fruit, and small arthropods, especially beetles, caterpillars, and spiders. Generally seen singly or in a pair, but family groups may stay together for a short time after the young fledge. Nests on ground, or occasionally in low trees or shrubs. Sings a simple trill of variable speed from an exposed perch, though geographical variations occur. Call is an upslurred, inquisitive *queee.*

Habitat

Common in brushy thickets and forest edges. Visits backyards where seed is scattered on the ground.

Local Sites

Red Rocks Park is an excellent place to see the Spotted Towhee all year.

FIELD NOTES The Canyon Towhee, *Pipilo fuscus* (inset), largely replaces the Spotted Towhee in the more open arid habitats of southeastern Colorado. It is gray above, paler on the underparts, and has a reddish crown sometimes raised into a crest. A dark central breast spot and fine streaks on the throat also characterize this inhabitant of semiarid desert canyons.

Nonbreeding | Adult

AMERICAN TREE SPARROW

Spizella arborea L 6.3" (16 cm)

FIELD MARKS

Gray head and nape crowned with rufous; rufous stripe behind eye

Gray throat, breast, with dark spot in center; rufous patches at sides of breast, gray-white underparts

Back streaked with black and rufous; notched tail

Behavior

Despite its name the American Tree Sparrow forages on the ground, nests on the ground, and breeds above the tree line in the far north. Eats insects during the breeding season, but eats chiefly seeds and plant matter during the winter. Gives a musical *teedle-eet* call, as well as a thin *seet*. Song begins with several clear notes followed by a variable, rapid warble.

Habitat

Although in general this sparrow likes open areas with scattered trees and brush, in the winter it prefers areas near humans, where seeds from bird feeders are plentiful.

Local Sites

In winter, watch backyard bird feeders for the American Tree Sparrow, or look for it foraging in shrubby areas throughout the state.

FIELD NOTES The American Tree Sparrow is the only member of the genus *Spizella* likely to be seen in Colorado during winter. Four other species belonging to this genus are found in Colorado, including three that breed in the state. All are small, slender birds with proportionately long, notched tails.

Breeding | Adult

CHIPPING SPARROW

Spizella passerina L 5.5" (14 cm)

FIELD MARKS

Streaked brown wings and back; unstreaked gray breast and belly; dark line through eye; gray rump

Breeding adult has chestnut crown; gray cheek and nape

Winter adult has streaked brown crown and a brown face

Behavior

Forages on the ground for insects, caterpillars, spiders, and seeds. Nests close to the ground in branches or vine tangles. May be found foraging in small family flocks in fall or in mixed-species groups in winter. Sings from high perch a one-pitched, rapid-fire trill of dry *chip* notes. Call in flight or when foraging is a high, hard *seep* or *tsik*.

Habitat

The Chipping Sparrow can be found in suburban lawns and gardens, woodland edges, and pine and oak forests. Frequents more open areas in winter.

Local Sites

Common summer resident in open coniferous forests and montane foothills throughout the major mountains of Colorado. After breeding, Chipping Sparrows disperse throughout the state from alpine tundra to grasslands. Most are gone by the end of November, and return in mid-April.

FIELD NOTES Brewer's Sparrow, *Spizella breweri* (inset), is a common breeding species in sagebrush habitat across much of Colorado, and a common migrant throughout the state in open habitats. Its nape and brown rump are streaked, and it has a distinctive white eye ring.

Year-round | Adult

VESPER SPARROW

Pooecetes graminius L 6.3" (16 cm)

FIELD MARKS

White outer tail feathers

White eye ring

Streaked upperparts, breast, flanks

Pale ear patches bordered by dark markings

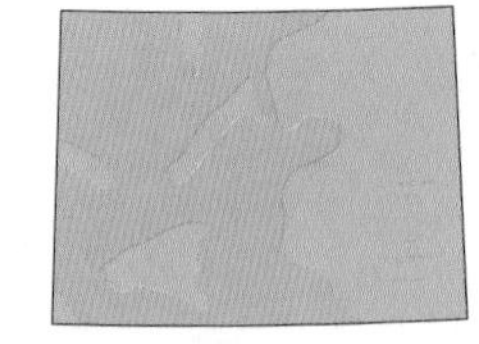

Behavior

Vesper Sparrows are more easily observed than many other grassland sparrows and are frequently seen along roadsides. When flushed, they often fly to the top of shrubs or fences. As the name suggests, Vesper Sparrows frequently sing in the evening (as well as early morning). The delightful song consists of doubled whistles and musical trills that quicken and descend in pitch.

Habitat

Found in a variety of dry open habitats, usually with a mixture of grass and shrubs. In Colorado, most breed in montane meadows, grasslands, and sagebrush steppe.

Local Sites

This fairly common breeding sparrow is best found by listening for its whistled song in open habitat at places like Rocky Mountain National Park, mostly from late April through September.

FIELD NOTES Another musical songster, the Cassin's Sparrow, *Aimophila cassini* (inset), is nearly impossible to detect when not singing. While it's most easily identified by song, also note the Cassin's unmarked breast, pale eye ring, indistinct supercilium, and blurry streaks on its rear flanks.

Breeding | Male

LARK BUNTING

Calamospiza melanocorys L 7" (18 cm)

FIELD MARKS

Heavy blue-gray bill

White or buff wing patches

Breeding male has black body; female is brown above, streaked below; winter male like female but blacker face and wings

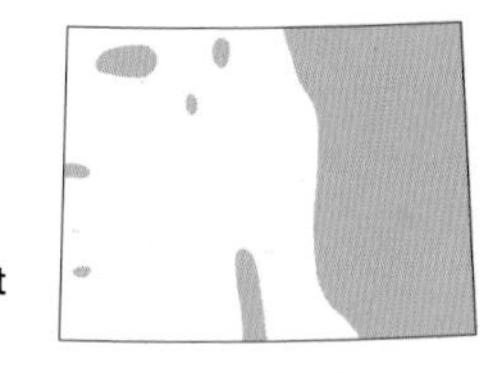

Behavior

The sociable Lark Bunting is often seen in large flocks, feeding on the ground, foraging for seeds and insects. Colorado's state bird actually spends little time in the state, most arriving in early May and leaving by early September. The distinctive call is a soft *hoo-ee,* and the song is a varied series of rich whistles and trills.

Habitat

In breeding season, favors valley floors, especially those with grasses and weeds, and a variety of other open habitats. Breeding habitat is mostly native short-grass prairie.

Local Sites

Lark Bunting populations fluctuate widely from year to year. It may be seen throughout eastern Colorado, but Pawnee National Grasslands is a particularly sensational place to watch these aerial songsters.

FIELD NOTES The attractive harlequin-faced Lark Sparrow, *Chondestes grammacus* (inset), also breeds in Colorado. It is found in summer in dry, open woodlands, usually with sandy soils.

Year-round | Adult

SONG SPARROW

Melospiza melodia L 5.8-7.5" (15-19 cm)

FIELD MARKS

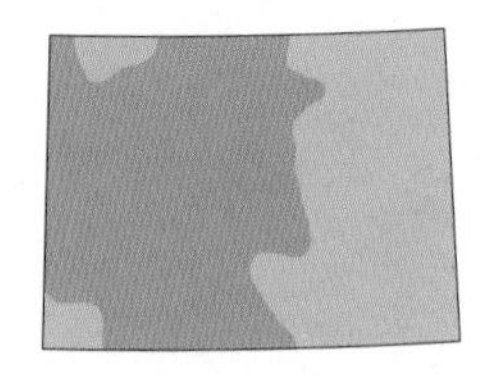

Underparts whitish, with streaks on sides and breast that converge into dark breast spot

Streaked brown and gray above; broad, grayish eyebrow; broad, dark malar stripe

Long, rounded tail

Behavior

Forages in trees and bushes and on the ground for larvae, fruits, and berries, sometimes scratching ground to unearth grain or insects. One of the most frequent hosts to cowbird parasitism, the Song Sparrow has learned to drive the menace away from nesting areas. Melodious song consists of three to four short, clear notes followed by a buzzy *tow-wee* and a trill. Distinctive call is a nasal, hollow *chimp.*

Habitat

Found year-round in a variety of habitats, including suburban and rural gardens, weedy fields, and forest edges. Favors dense streamside thickets for breeding.

Local Sites

Commonly seen year-round in moist, brushy habitats along the Front Range.

FIELD NOTES Spreading across most of North America, from the Aleutian Islands of Alaska to the borders of Mexico and eastward along the Atlantic Coast, there are over 30 recognized subspecies of Song Sparrow, all of which have adapted to various specific environments. Pale races, such as *saltonis,* inhabit arid regions in the Southwest; darker races, such as the eastern *melodia,* inhabit more humid regions; and larger races, such as the Alaskan *maxima,* inhabit northern islands.

Year-round | Adult

LINCOLN'S SPARROW

Melospiza lincolnii L 5.8" (15 cm)

FIELD MARKS

Finely streaked breast, washed with warm buff

Buffy malar stripe and eye ring

Small slender bill

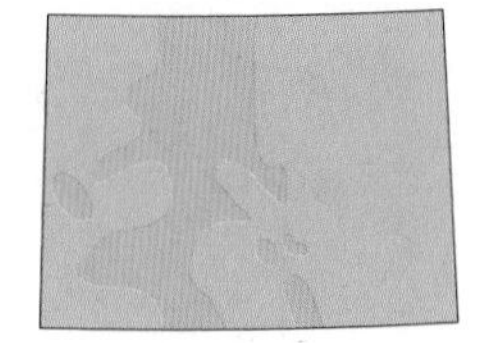

Behavior

Lincoln's Sparrows generally skulk in moist or shrubby vegetation, typically below eye level. They are rarely seen in the open. The rich warbled and bubbling song is reminiscent of the House Wren, but with a more liquid tone.

Habitat

In Colorado breeds exclusively in montane areas, in bogs, wet meadows, and riparian thickets. In migration and winter found in brushy areas, thickets, hedgerows, forest edges, clearings, and shrubby areas. Rarely detected in winter in Colorado.

Local Sites

Moist thickets at Rocky Mountain National Park, especially at Endovalley Campground, are reliable locations for finding Lincoln's Sparrow in summer. It is a common breeder in appropriate habitat, and it should be easily found by listening for its song.

FIELD NOTES Similar in plumage and overall coloration to the larger Song Sparrow, *Melospiza melodia* (p. 218), the Lincoln's Sparrow is most easily identified by its thinner breast streaks, thin, buffy malar stripe, and prominent eye ring. The Lincoln's Sparrow shows very little variation in plumage throughout its extensive range.

Year-round | Adult "Gambel's"

WHITE-CROWNED SPARROW

Zonotrichia leucophrys L 7" (18 cm)

FIELD MARKS

Black-and-white striped crown

Underparts mostly gray

Brownish upperparts with both pale and black streaks

Pink, orange, or yellowish bill—depending on subspecies

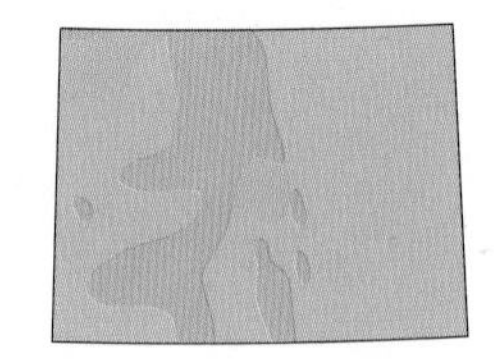

Behavior

Scratches feet along the ground, foraging for insects, caterpillars, and seeds. The operation is audible in areas where wintering flocks congregate. Also gleans food from vegetation. Song variable by region and often heard in winter. Usually one or more thin, whistled notes followed by a twittering trill. Calls include a loud *pink* and a sharp *tseep*.

Habitat

The White-crowned Sparrow occurs in woodlands, grasslands, roadside hedges. Nests on piles of grass or moss, usually in a bush or tree.

Local Sites

Very common to locally abundant migrant and winter resident. Common to abundant breeder at high elevations near timberline.

FIELD NOTES The common winter subspecies group of White-crowned, "Gambel's," has whitish supraloral areas and an orange bill. The "Rocky Mountain" group that breeds in Colorado has a dark supraloral area and a dark pink bill.

Year-round | Adult "Gray-headed"

DARK-EYED JUNCO

Junco hyemalis L 6.3" (16 cm)

FIELD MARKS

"Gray-headed" described below

Mostly gray plumage; white outer tail feathers visible in flight; rufous back

Pink bill; black face; dark eyes

Behavior

Scratches on ground and forages by gleaning seeds, grain, berries, insects, caterpillars, and fruit from plants. Occasionally gives chase to a flying insect. Forms flocks in winter, when males may remain farther north or at greater elevations than juveniles and females. Song is a short, musical trill that varies in pitch and tempo. Calls include a sharp *dit,* and a rapid twittering in flight.

Habitat

Summers in coniferous forests. Winters in a variety of habitats, often in patchy wooded areas.

Local Sites

The "Gray-headed" subspecies group is a common summer resident in coniferous forests at higher elevations in most of the region's northern mountains. Dark-eyed Juncos are common throughout the state in winter.

FIELD NOTES All five subspecies groups of the Dark-eyed Junco can be seen in Colorado. The breeding "Gray-headed" juncos are augmented by "Slate-colored" (inset, top), and "Oregon" (inset, bottom), as well as the "Pink-sided" and "White-winged."

Breeding | Adult male

CHESTNUT-COLLARED LONGSPUR

Calcarius ornatus L 6" (15 cm)

FIELD MARKS

Breeding male has rufous nape; black belly; yellow ear patch

Breeding female is subdued version of breeding male

Small bill; short primary projection

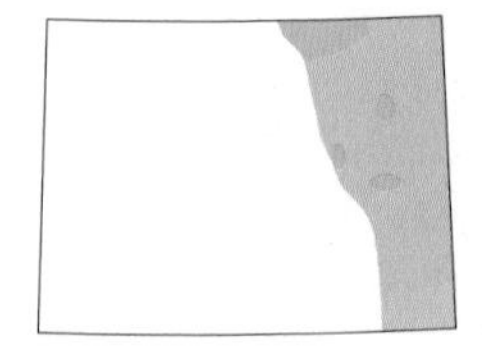

Behavior

Outside the breeding season, Chestnut-collared Longspurs are almost always found on the ground, where they forage for seeds and insects. The short musical song sounds similar to the song of a Western Meadowlark (p. 235). At other seasons, Chestnut-collared Longspurs are most frequently detected by their dry mechanical rattle calls.

Habitat

Found in short-grass prairie, generally preferring areas with a richer variety of grass and shrubs in Colorado. During migration and winter, may be found in agricultural areas. Historically bred at sites recently grazed by bison or disturbed by fire.

Local Sites

The Pawnee National Grassland is the best place to see this stunning songster. It is most easily seen from April through June when males are performing their aerial courtship displays.

FIELD NOTES Colorado's other breeding longspur, the McCown's Longspur, *Calcarius mccowni* (inset, breeding: male, top; female, bottom) has a larger bill and unsteaked breast. Also note the patch of cinnamon on adults' wings. They favor shorter grass than Chestnut-collareds.

Breeding | Adult male

BLACK-HEADED GROSBEAK

Pheucticus melanocephalus L 8.3" (21 cm)

FIELD MARKS

Male has cinnamon underparts and nape; black head and upperparts; two prominent wing bars

Female buffy overall, with streaked mantle

Yellow wing linings show in flight; large, dark, triangular bill

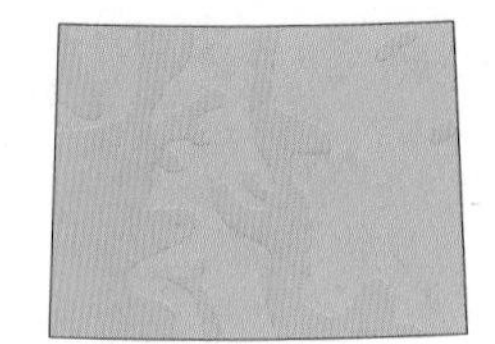

Behavior

Forages for seeds, insects, caterpillars, berries, and fruit on the ground and in trees and bushes. Generally seen singly or in a pair. Nests moderately high up in dense vegetation of trees or shrubs, usually near water. Nestlings are brooded by both male and female. Song is a rich, whistled warble. Very soft songs may also be emitted by either sex while incubating eggs. Call is a sharp *eek*.

Habitat

Inhabits open and streamside woodlands and forest edges. Known to visit backyard feeders and suburban parks.

Local Sites

In summer, listen for the Black-headed's melodious song in deciduous and mixed forests throughout Colorado, especially in riparian areas and foothills.

FIELD NOTES The Black-headed Grosbeak will hybridize with its cousin, the Rose-breasted Grosbeak, *Pheucticus ludovicianus,* where their ranges overlap in the Great Plains. Their nearly identical songs no doubt promote the interchange. Hybrids may occasionally occur at feeders in Colorado. Males are marked by underparts mottled in red, orange, and white and by a dusky orange and black head pattern.

Year-round | Adult male

LAZULI BUNTING

Passerina amoena L 5.5" (14 cm)

FIELD MARKS

Male has bright blue hood and upperparts; cinnamon upper breast; white belly and undertail coverts; two white wing bars

Female drab brown overall, with bluish rump and tail; whitish belly; two faint buffy wing bars

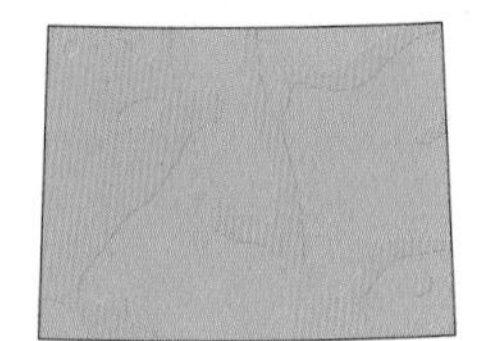

Behavior

Forages on ground and in low foliage primarily for seeds. May also consume insects and caterpillars. Generally seen singly or in a pair, but may join small flocks after breeding and larger flocks during migration, sometimes with sparrows or other species of buntings. Nests close to ground in small bush or tree. Persistently sings a vivacious series of varied phrases, sometimes with repeated notes. Highly territorial, many young males learn songs not from parents, but from competing males. Call is a short *pik*.

Habitat

Found in a variety of low-elevation brushy habitats, especially on hillsides, in valleys, and along streams.

Local Sites

Most widespread west of the Rocky Mountains. Also found along the Front Range at places like Red Rocks Park.

FIELD NOTES The female Lazuli Bunting (inset) has drab brown plumage overall in stark contrast to the male's sky blue hood and back. At first glance, she can be mistaken for a sparrow, but note her unstreaked back, warm buffy breast, and the blue tint to her rump, tail coverts, and to a lesser extent her wing coverts.

Year-round | Adult male

RED-WINGED BLACKBIRD

Agelaius phoeniceus L 8.7" (22 cm)

FIELD MARKS

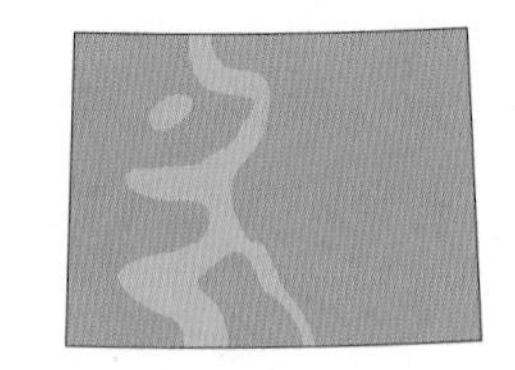

Male is glossy black with bright red shoulder patches broadly edged in buffy yellow

Females densely streaked overall

Pointed black bill

Wings slightly rounded at tips

Behavior

Runs and hops while foraging for insects, seeds, and grains in pastures and open fields. The male's bright red shoulder patches are usually visible when it sings from a perch, often atop a cattail or tall weed stalk, defending its territory. At other times only the yellow border may be visible. Territorially aggressive, a male's social status is dependent on the amount of red he displays on his shoulders. Song is a liquid, gurgling *konk-la-reee,* ending in a trill. Call is a low *chack* note.

Habitat

Breeds in colonies, mainly in freshwater marshes and wet fields with thick vegetation. Nests in cattails, bushes, or dense grass near water. During winter, flocks forage in wooded swamps and farm fields.

Local Sites

The Red-winged, common to abundant year-round in wetlands throughout Colorado, forms large flocks in fall and winter.

FIELD NOTES Usually less visible within large breeding colonies, the female Red-winged (inset) is streaked dark brown above and has dusky white underparts heavily streaked with dark brown. In winter, whole flocks may form of females only.

Year-round | Adult

WESTERN MEADOWLARK

Sturnella neglecta L 9.5" (24 cm)

FIELD MARKS

Black V-shaped breast band on yellow underparts

Long, pointed bill

Brown and white crown stripes; black and brown barred back

Heavily streaked white flanks

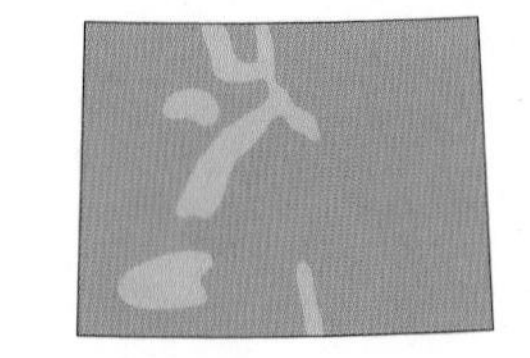

Behavior

A ground feeder with strong legs for walking; forages through marsh edges, lakeshores, fields, meadows, and lawns, gleaning whatever might be available. Uses long, thin, sharp-tipped bill to probe deep into soil to pluck out a variety of food including seeds, fruit, insects, and worms. Found singly or in a pair during breeding season, but highly gregarious at other times of the year.

Habitat

Prefers the open space offered by pastures, grasslands, and farm fields. Nests on the ground by creating a dome or partial dome of grass and weeds. Often perches on fence posts and utility wires.

Local Sites

A year-round resident of Colorado, large flocks often gather and perch on fence posts near roadsides. Easily seen at Pawnee and Comanche National Grasslands.

FIELD NOTES Sings from exposed perch a series of bubbling, melodious notes of variable length, usually accelerating toward the end. Gives a sharp chuck note on the ground and a whistled *wheet* call in flight.

Breeding | Adult male

YELLOW-HEADED BLACKBIRD

Xanthocephalus xanthocephalus L 9.5" (24 cm)

FIELD MARKS

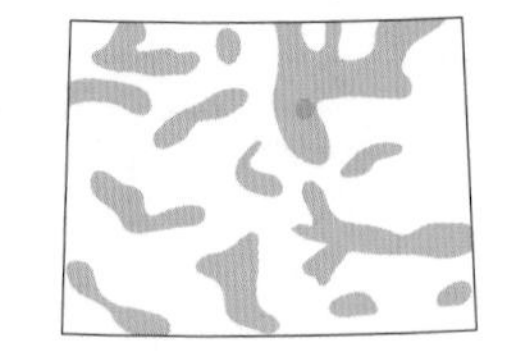

Male has prominent yellow hood and breast; black body and lores; large white wing patch

Female is washed with yellow on her face and breast; dark brown body and crown

Stubby, black, triangular bill

Behavior

Highly gregarious, the Yellow-headed Blackbird breeds in colonies and forms large flocks outside of the breeding season, sometimes numbering in the thousands. Forages communally on ground for insects, larvae, snails, grain, and seeds. Highly territorial, the Yellow-headed will attack other birds and even humans who intrude on its territory. Unmusical, raspy song begins with a few loud, harsh notes and ends in a long, descending buzz resembling the sound of a chainsaw. Call note is a hoarse croak.

Habitat

Found at freshwater bodies of water such as marshes, reedy lakes, and cattail swamps. Nests among grasses and reeds just above the water's surface.

Local Sites

Barr Lake State Park, the San Luis Valley, and most major marshes host breeding colonies of Yellow-headed Blackbirds in summer.

FIELD NOTES The female Yellow-headed weaves her nest of sedges and grass in emergent vegetation in or near water. She seems to intentionally use wet materials in the construction of her nest, and as the wet vegetation dries, it shrinks and tightens the weave into a very sturdy structure.

Year-round | Adult male

BREWER'S BLACKBIRD

Euphagus cyanocephalus L 9" (23 cm)

FIELD MARKS

Male is black with purplish gloss on head and neck; greenish gloss on body and wings; less glossy in winter

Male has yellow eyes; female has brown eyes, gray-brown body

Bill is sharp and straight

Behavior

Like all blackbirds, the Brewer's has strong legs and feet that allow it to walk for long stretches as it forages on the ground. With straight, strong, sharp-tipped bill, it gleans insects, fruit, grain, and seeds. Raises tail and inclines body while foraging. Will assemble in parking lots to pick protein-rich insects from car grilles and to scavenge handouts. Female builds coarse, cup-shaped nest of needles, grasses, and twigs; pads the inside of the cup with either mud or cow manure. Spreads its tail and droops its wings as it sings its short, raucous, wheezy *quee-ee* or *k-seee*. Call is a harsh *check*.

Habitat

Common in open habitats, including fields, marshes, suburbs with parks, and parking lots. Breeds away from cities in agricultural areas or grasslands.

Local Sites

Check the nearest supermarket parking lot in the foothill and mountain towns for this ubiquitous year-round resident.

FIELD NOTES Strong jaw musculature allows blackbirds to close their bills—and then forcefully open them in an action called "gaping." Gaping allows the birds to pry into crevices, soft bark, dirt, and leaf litter to expose prey unavailable to other birds.

Year-round | Adult

COMMON GRACKLE

Quiscalus quiscula L 12.6" (32 cm)

FIELD MARKS

Plumage appears all black; in good light, shows glossy blue hood, bronze body, purple tail

Long, wedge-shaped tail

Pale yellow eyes

Female plumage not as iridescent

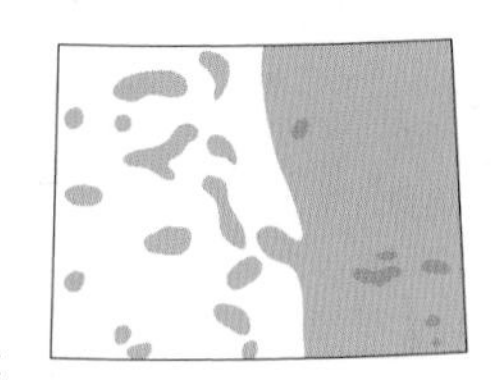

Behavior

Rarely seen outside of a flock, this grackle moves to large, noisy, communal roosts in the evening. During the day, mainly seen on the ground in a group, feeding on insects, spiders, grubs, and earthworms. Sometimes wades into shallow water to forage for minnows and crayfish; known to feed on eggs and baby birds. Courtship display consists of male puffing out shoulder feathers to make a collar, drooping his wings, and singing. The Common Grackle produces a sound like ripping cloth or cracking twigs. Call note is a loud *chuck.*

Habitat

Prefers open spaces provided by farm fields, pastures, marshes, and suburban yards. Requires wooded areas, especially conifers, for nesting and roosting.

Local Sites

Abundant throughout eastern portions of Colorado, the Common Grackle has increasingly expanded westward into urban areas since the 1950s.

FIELD NOTES Even larger and more vocal is the Great-tailed Grackle, *Quiscalus mexicanus* (inset: male), a species that is rapidly colonizing the state.

Year-round | Adult male

BROWN-HEADED COWBIRD

Molothrus ater L 7.5" (19 cm)

FIELD MARKS

Male's brown head contrasts with metallic black body

Female gray-brown above, paler below with a whitish throat

Short, dark, pointed bill

Juveniles streaked below

Behavior

Often forages on the ground among herds of cattle, feeding on insects flushed by the grazing animals. Also feeds heavily on grass seeds and agricultural grain, and is sometimes viewed as a threat to crops. Generally cocks its tail up while feeding. The Brown-headed Cowbird is a nest parasite and lays its eggs in the nests of other species, leaving the responsibilities of feeding and fledging of young to the host birds. Song is a squeaky gurgling. Calls include a squeaky whistle.

Habitat

Cowbirds prefer open habitat such as farmlands, pastures, prairies, and edgelands bordering forests. Also found around human habitation.

Local Sites

Brown-headed Cowbirds are common throughout Colorado, in relatively open and fragmented habitats.

FIELD NOTES The Brown-headed Cowbird flourishes throughout North America, adapting to newly cleared lands and exposing new songbirds—now more than 200 species—to its parasitic brooding habit. The female Brown-headed Cowbird lays up to 40 eggs a season in the nests of host birds, leaving the task of raising their young to the host species.

Year-round | Adult male

BULLOCK'S ORIOLE

Icterus bullockii L 8.7" (22 cm)

FIELD MARKS

Male has bold orange face and underparts; black crown, back, tail, throat patch, and eye line; large white wing patches

Female has yellow face, throat, and breast; drab olive wings, back, and tail; grayish belly

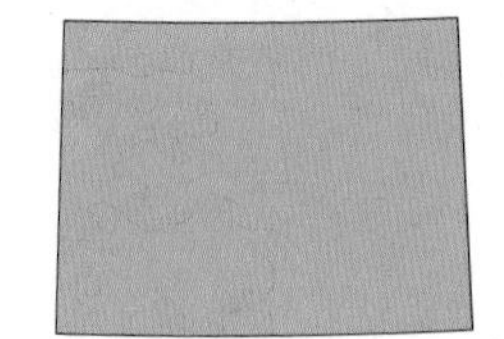

Behavior

Forages in trees and bushes for insects, berries, and fruit. Inserts long, sharply pointed bill into crevices to probe for ants, mayflies, and spiders. In breeding season, male chases female with such actions as wing-drooping and repeated bowing, displaying his brilliant orange plumage. Pairs are noisy and conspicuous, and spend much time together, but mate for only one season. Female weaves grasses into intricate hanging baskets or pouches for nests. Song is variable, but always composed of whistles and harsher notes; call is a clear, harsh *cheh,* sometimes given in a series.

Habitat

Breeds in open wooded areas, especially those rife with deciduous trees.

Local Sites

Look for this striking bird by itself or in a small family group in summer at Barr Lake State Park or other locations with riparian habitat in Colorado.

FIELD NOTES The Bullock's Oriole was once considered the same species as the Baltimore Oriole, *Icterus galbula* (inset: adult male), which breeds in eastern Colorado. The two hybridize where their ranges overlap in the Great Plains, though the male Baltimore has a full black hood and much less white in his wings. Female Baltimores are more orange than female Bullock's.

Year-round | Adult

BROWN-CAPPED ROSY-FINCH

Leucosticte aistralis L 6" (15 cm)

FIELD MARKS

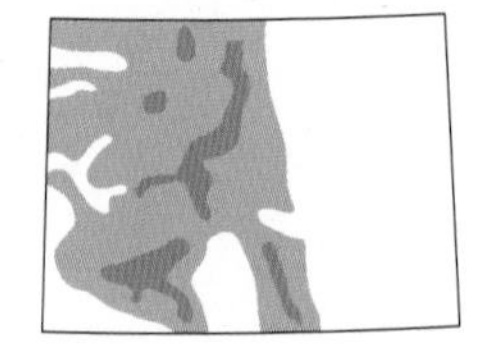

Brown face; limited gray on dark centered crown; pinkish belly

Black bill in summer, yellowish with black tip in winter.

First-winter birds more dully marked; pink coloration often limited to wings.

Behavior

Foraging rosy-finches pick insects, spiders, and seeds from the surface of snow on alpine tundra. The tightly woven cup nest is placed under large rocks on cliff faces or rockslides, infrequently in abandoned mines and railroad tunnels. The rarely heard song consists of a series of whistled notes. The buzzy husky call notes are similar to those of the House Sparrow (p. 260) or Evening Grosbeak (p. 258).

Habitat

Found exclusively on alpine tundra during the breeding season, where they nest on rocky cliffs and forage for insects and seeds along edges of icefields.

Local Sites

During the summer, look for this species in rocky areas with cliffs above timberline at locations such as Mount Evans and Rocky Mountain National Park. Brown-capped Rosy-Finches may also be found at mountain feeders in the winter.

FIELD NOTES During winter, Brown-capped Rosy-Finches are often joined by Black Rosy-Finches, *Leucosticte atrata* (inset, breeding: female, top; male, bottom), and Gray-crowned Rosy-Finches, *Leucosticte tephrocotis*. They are most easily found on snowy days at mountain bird feeders.

Year-round | Adult male

CASSIN'S FINCH

Carpodacus cassinii L 6.3" (16 cm)

FIELD MARKS

Underparts with crisp streaking, including undertail coverts

Pale eye ring; pale grayish brown upperparts with crisp blackish streaking

Adult male: rosy pink wash on head, breast, uppertail coverts

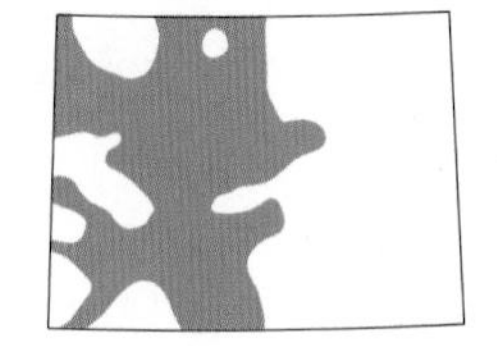

Behavior

The Cassin's Finch often breeds in small colonies, with nests sometimes as close as three feet apart. If the nests are this close, the males usually fight until one of the pair gives up. If the first nest is started substantially earlier than the other, however, such close nesting may be tolerated. The rich warbled song of the Cassin's Finch often includes imitations of other birds.

Habitat

Found in a variety of coniferous habitats in the mountains, but also frequently seen feeding in deciduous trees. A frequent visitor to bird feeders.

Local Sites

This familiar year-round visitor to mountain bird feeders can be found in most mountainous areas in the state. Rocky Mountain and Genesee National Parks are two of the most reliable locations for finding Cassin's Finches near the Front Range.

FIELD NOTES Cassin's Finches take a year to acquire adult plumage, making it impossible to differentiate immature males from immature females and adult females in the field. This may lead to the mistaken impression that both sexes regularly sing, not just the male.

Year-round | Adult male

HOUSE FINCH

Carpodacus mexicanus L 6" (15 cm)

FIELD MARKS

Male's forehead, bib, and rump typically red, but can be orange or, occasionally, yellow

Brown streaked back; white belly; streaked flanks

Female streaked dusky brown on entire body

Behavior

A seed eater, the House Finch forages on the ground, in fields and in suburban yards. Often visits backyard feeders. Seen in large flocks during winter. Flies in undulating pattern, during which squared-off tail is evident. Male sings a conspicuously lively, high-pitched song consisting of varied three-note phrases, usually ending in a nasal *wheer.* Calls include a whistled *wheat.*

Habitat

Adaptable to varied habitats, this abundant bird prefers open areas, including suburban parks and areas where it can build its cuplike nest on buildings. Also nests in shrubs, trees, cactuses, or on the ground.

Local Sites

The House Finch is fairly common year-round in much of Colorado, particularly in towns and suburbs.

FIELD NOTES The female House Finch (inset) is grayish brown overall and heavily streaked on her entire body. Pairs can often be found during breeding season, and small family groups after nesting, but this gregarious bird forms large foraging flocks for the winter, sometimes with other species.

Year-round | Adult

PINE SISKIN

Carduelis pinus L 5" (13 cm)

FIELD MARKS

Streaked brown overall; short tail

Flight feathers and tail are washed with yellow, more evident on male

Two wing bars

Thin pointed bill

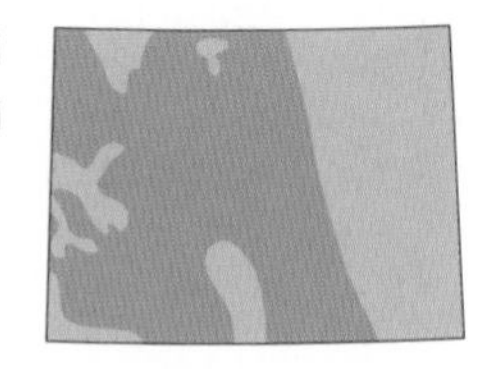

Behavior

Almost always seen in flocks, foraging on the ground or in trees for seeds and, in summer, some insects. Often found in fields of thistle in the company of goldfinches. May also drink sap from tree wells drilled by sapsuckers. A highly nomadic species. Nest generally located far out on a branch of a conifer. Song is a twittering, variably pitched, jumbled warble. Call is an ascending, buzzy *zreee.* Flight call is a hoarse, repeated *chee.*

Habitat

Found mostly in coniferous or mixed woodlands. Erratic, unpredictable movement of flocks brings these birds also into urban parks and weedy fields. Will visit goldfinch feeders.

Local Sites

Siskins breed in the coniferous forests of all of Colorado's mountain ranges. In winter, they will descend to nearby foothills, valleys, and plains.

FIELD NOTES Another permanent resident of Colorado is the Red Crossbill, *Loxia curvirostra* (inset: female, top; male, bottom). Easily distinguished from siskins by overall body color, crossbills are distinctive in that their mandibles cross at the tips, enabling these birds to extract seeds with ease from the cones of coniferous trees.

Year-round | Adult male "Black-backed"

LESSER GOLDFINCH

Carduelis psaltria L 4.5" (11 cm)

FIELD MARKS

Male black on hood, back, and tail; bright yellow below

Female greenish olive above, dull yellow below

White wing bars and edges to tertial and primary flight feathers

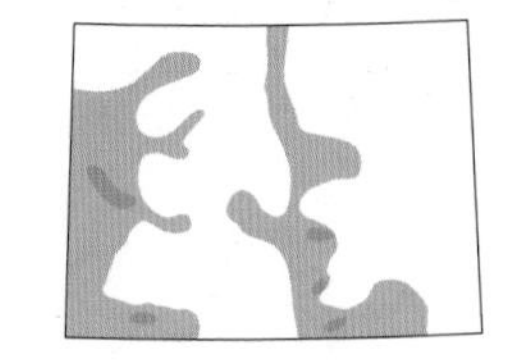

Behavior

Pairs or small flocks forage for seeds and insects in bushes, shrubs, and weedy fields. Commonly visits birdbaths and outdoor faucets in semiarid areas, as diet of primarily seeds does not provide a lot of moisture. Female builds nest in bushes or trees, sometimes in tall weeds. Male feeds female while she is brooding and pairs may stay together for life. Song, given by male in flight, is a lively series of warbles and *swee* notes. Call, often given by small flocks, is a *tee-yee tee-yer.*

Habitat

Found in dry brushlands and open woodlands with scattered trees. Also tends to areas of human habitation to take advantage of artificial sources of water.

Local Sites

Lesser Goldfinches nest along the length of the Front Range and most of western Colorado.

FIELD NOTES Colorado populations of Lesser Goldfinches include males with uniformly black backs (opposite), and males with green upperparts (inset: male adult, top; immature male, below). Most depart the state in winter, but some are regular winter visitors, particularly in the Grand Junction area.

Breeding | Adult male

AMERICAN GOLDFINCH

Carduelis tristis L 5" (13 cm)

FIELD MARKS

Breeding male bright yellow with black cap; female and winter male duller overall, lacking cap

Black wings have white bars

Black-and-white tail; white undertail coverts

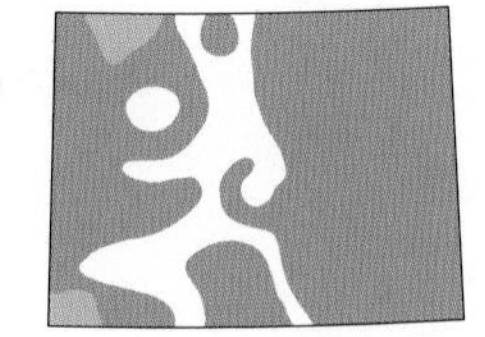

Behavior

Gregarious and active. Winter flocks may contain a hundred or more goldfinches and include several other species. The typical goldfinch diet, mostly seeds, is the most vegetarian of any North American bird; the goldfinch, however, sometimes eats insects as well. During courtship, male performs exaggerated, undulating aerial maneuvers, and often feeds the incubating female. Song is a lively series of trills, twitters, and *swee* notes. Distinctive flight call is *per-chik-o-ree.*

Habitat

Common but declining in weedy fields, open second-growth woodlands, and anywhere rich in thistles and sunflowers. Nests at edges of open areas or in old fields, often late in summer after thistles have bloomed so that the soft parts of the plant can be used as nest lining.

Local Sites

Commonly seen throughout Colorado. Often visits seed feeders, particularly those offering "thistle" seed.

FIELD NOTES Winter-plumaged males (inset) and females are more subtly attired, but still have bold wing bars and white undertail coverts.

Year-round | Adult male

EVENING GROSBEAK

Coccothraustes vespertinus L 8" (20 cm)

FIELD MARKS

Stocky finch with large, pale yellow or greenish bill

Yellow eyebrow and forehead on adult male; dark brown and yellow body; white secondaries

Gray and tan female has thin, dark malar stripe; white-tipped tail

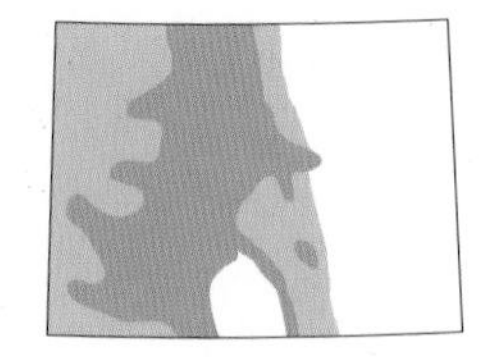

Behavior

Forages mostly in trees and shrubs for seeds, berries, and insects, sometimes searching on the ground. Also feeds on buds of deciduous trees, and maple sap. At bird feeders, the Evening Grosbeak is fond of sunflower seeds, and uses its powerful jaws to crack them open easily. Gregarious year-round, they travel in flocks and even build their nests near each other. Loud, strident calls include a *clee-ip* and a *peeer*. Song consists of regularly repeated call notes.

Habitat

Breeds in montane coniferous and mixed woods. In winter, descends to lower-elevation woodlots, shade trees, and feeders.

Local Sites

Evening Grosbeaks can be found in summer in coniferous forests in all of Colorado's mountain ranges. The town of Allenspark, below Rocky Mountain National Park, is a particularly good spot year-round.

FIELD NOTES A fairly common visitor to backyard feeders, especially in winter, the Evening Grosbeak's (inset: female) pattern of occurrence may even be affected by the location and abundance of feeders. Grosbeaks may appear in one area one winter and be absent the next.

Summer | Adult male

HOUSE SPARROW

Passer domesticus L 6.3" (16 cm)

FIELD MARKS

Summer male has black bill, bib, and lores; chestnut eye stripes and nape

Winter male has chestnut and black areas veiled with gray

Female brown with streaked back; buffy eye stripe

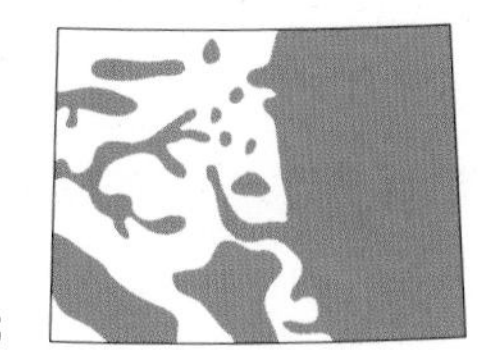

Behavior

The House Sparrow is abundant and gregarious year-round. Hops around, feeding on grain, seeds, and shoots, or seeks out bird feeders for sunflower seeds and millet. In urban areas, begs for food from humans and will clean up any crumbs left behind. In spring and summer, multiple suitors will chase a possible mate. Females choose mate mostly according to song display. Singing males give persistent *cheep*.

Habitat

Found in close proximity to humans. Can be observed in urban and suburban areas and in rural landscapes inhabited by humans and livestock. Nests in any sheltered cavity, often usurping it from another species.

Local Sites

Common to abundant permanent resident in towns and cities across Colorado.

FIELD NOTES These Old World sparrows were originally introduced into eastern North America and have expanded to virtually the entire country. The female (inset) is much plainer than the male; juveniles look similar to the female.

Color categories reflect the overall colors of a species, not just the head color. Where sexes or ages differ, we typically show the most colorful plumage.

Mostly Black

Mostly Black and White

Mostly Blue

Mostly Brown

Mostly Brown and White

Mostly Gray

Mostly Green

Mostly White

Prominent Yellow

Prominent Green Head

Prominent Orange

Prominent Red

The main entry page number for each species is listed in **boldface** type and refers to the text page opposite the illustration.

A check-off box is provided next to each common-name entry, so that you can use this index as a checklist of the species you have identified.

ACKNOWLEDGMENTS

The Book Division would like to thank the following people for their guidance and contribution in creating the *National Geographic Field Guide to Birds: Colorado.*

Cortez C. Austin, Jr.
Cortez Austin is a wildlife photographer specializing in North American and tropical birds. His photographs have appeared in field guides, books, and brochures on wildlife. He is also an ardent conservationist.

Richard Crossley
Richard Crossley is an Englishman obsessed by birding since age 10. He traveled the world studying birds but fell in love with Cape May while pioneering the identification of overhead warbler migration in 1985. He is co-author of *The Shorebird Guide,* due in Spring 2006.

Larry Sansone
Larry Sansone's pictures have been published worldwide. He was a technical advisor to the *National Geographic Field Guide to the Birds of North America* first edition and is photo editor of *Rare Birds of California* , by the California Bird Records Committee.

Brian E. Small
Brian E. Small has been a professional wildlife photographer specializing in birds for many years. He has been a columnist and Advisory Board member for *WildBird* magazine. Brian is currently Photo Editor for the American Birding Association's *Birding* magazine. You can find more of his images at www.briansmallphoto.com.

Bob Steele
Bob Steele has been involved in birding and bird photography for over 20 years. He lives in bird-rich Kern County, California. This area is centrally located at the convergence of multiple bio-regions, giving him a unique perspective on his avian subjects.

Brian Sullivan
Birding travels and field projects have taken Brian Sullivan to every corner of the globe. Research interests include migration, seabirds, raptors and bird identification. He is currently a PRBO Field Coordinator for the endangered San Clemente Loggerhead Shrike Recovery Project.

Tom Vezo
Tom Vezo is an award-winning wildlife photographer who is widely published throughout the U.S. and Europe. He is a contributor to the *National Geographic Reference Atlas to the Birds of North America.* Please visit Tom at his website www.tomvezo.com.

ILLUSTRATIONS CREDITS

Cortez C. Austin, Jr.: pp. 32, 48, 74, 150, 188. **Lance Beeny/VIREO:** p. 34. **Richard Crossley:** p. 182. **G.C. Kelley**: pp. 100, 108. **Larry Sansone:** pp. 90, 96, 120, 230. **Bill Schmoker:** p. 176. **Rulon E. Simmons:** p 210. **Brian E. Small:** pp. 28, 30, 36, 40, 54, 56, 70, 76, 80, 84, 94, 98, 104, 106, 112, 114, 118, 122, 126, 130, 138, 142, 144, 146, 160, 162, 164, 168, 174, 178, 184, 186, 192, 194, 198, 202, 204, 206, 208, 212, 216, 218, 226, 232, 236, 238, 244, 246, 248, 254, 256, 258. **Bob Steele:** cover, pp. 60, 68, 82, 92, 128, 134, 154, 190, 196, 214. **Brian Sullivan:** p. 124. **TomVezo.com:** pp. 14, 16, 18, 20, 22, 24, 26, 42, 44, 46, 50, 52, 58, 62, 64, 72, 78, 86, 88, 102, 110, 116, 132, 140, 148, 152, 156, 158, 166, 170, 172, 180, 200, 220, 222, 224, 228, 234, 240, 242, 250, 252, 260. **Brian K. Wheeler:** p. 66. **Christopher L. Wood:** pp. 2, 38, 136

NATIONAL GEOGRAPHIC
FIELD GUIDE TO BIRDS:
COLORADO

Edited by Jonathan Alderfer

**Published by
the National Geographic Society**

John M. Fahey, Jr.,
President and Chief Executive Officer

Gilbert M. Grosvenor,
Chairman of the Board

Nina D. Hoffman,
*Executive Vice President;
President, Books & School Publishing*

Prepared by the Book Division

Kevin Mulroy,
Senior Vice President and Publisher

Kristin Hanneman, *Illustrations Director*

Marianne R. Koszorus, *Design Director*

Carl Mehler, *Director of Maps*

Barbara Brownell Grogan,
Executive Editor

Staff for this Book

Barbara Levitt, *Editor*

Kate Griffin, *Illustrations Editor*

Carol Norton, *Series Art Director*

Alexandra Littlehales, *Designer*

Suzanne Poole, *Text Editor*

Teresa Neva Tate, *Illustrations Specialist*

Paul Hess, *Map Researcher*

Matt Chwastyk, Sven M. Dolling,
Map Production

Lauren Pruneski, Michael Greninger,
Editorial Assistants

Rick Wain, *Production Project Manager*

Manufacturing and Quality Control

Christopher A. Liedel,
Chief Financial Officer

Phillip L. Schlosser, *Vice President*

John T. Dunn, *Technical Director*

One of the world's largest nonprofit scientific and educational organizations, the National Geographic Society was founded in 1888 "for the increase and diffusion of geographic knowledge." Fulfilling this mission, the Society educates and inspires millions every day through its magazines, books, television programs, videos, maps and atlases, research grants, the National Geographic Bee, teacher workshops, and innovative classroom materials. The Society is supported through membership dues, charitable gifts, and income from the sale of its educational products. This support is vital to National Geographic's mission to increase global understanding and promote conservation of our planet through exploration, research, and education.

For more information, please call
1-800-NGS LINE (647-5463) or write
to the following address:

National Geographic Society
1145 17th Street N.W.
Washington, D.C. 20036-4688 U.S.A.

Log on to nationalgeographic.com;
AOL Keyword: NatGeo.

For information about special discounts
for bulk purchases, please contact
National Geographic Books Special Sales:

ngspecsales@ngs.org

**Library of Congress
Cataloging-in-Publication Data**

Available upon request

ISBN-10: 0-7922-5561-5

ISBN-13: 978-0-7922-5561-1